FLY ON INSTRUMENTS

The Pilot's Library Series

The Pilot's Library

General Editor, Robert B. Parke

FLYING AIRPLANES: THE FIRST HUNDRED HOURS by Peter Garrison
FLY ON INSTRUMENTS by George C. Larson

In Preparation:

WEATHER REPORT by Robert B. Parke
PILOT'S AVIONICS HANDBOOK by George C. Larson and Robert Denny
LONG DISTANCE FLYING by Peter Garrison
PILOT'S GUIDE TO PREVENTIVE AIRCRAFT MAINTENANCE by J. Mac McClellan

Other Doubleday Aviation Handbooks:

ANYONE CAN FLY by Jules Bergman
THE PILOT'S NIGHT FLYING HANDBOOK by Len Buckwalter
THE PILOT'S GUIDE TO FLIGHT EMERGENCY PROCEDURES by Alan Bramson
    and Nellie Birch

A PILOT'S LIBRARY BOOK

FLY ON INSTRUMENTS

by George C. Larson

With an Introduction by Robert B. Parke

COLUMBUS BOOKS

ISBN: 0-385-14619-1

Library of Congress Catalog Card Number 79-7602

Copyright © 1980 by George C. Larson
Printed in the United States of America

First Edition

CONTENTS

ACKNOWLEDGMENTS

Robert Parke persisted, nagged, cajoled, and eventually persuaded me to write "some personal thoughts on instrument flying." Sounded modest enough, and I doubt that either one of us knew what we were in for; but both of us know now. As my editor when we worked together at *Flying* magazine, his consummate talents were persuasion coupled with an uncommon vision. Had he not seen the need for this book, it would never have been done.

During the time I grew up within aviation, I was fortunate to be surrounded by people who talked about it, loved it, made fun of it, and thought deeply about it and at great length. Revolving around the sun of *Flying*'s New York office were some planetary personalities who generated an intelligence and an energy that was truly synergistic; we all felt privileged to be part of what, at the time, seemed an epoch. They're all woven into this book's cloth, and although they kept me thinking at a dead run for more than five years, it was so damned much fun I forgot to breathe very hard.

A couple of years ago, I took an instrument refresher course at Aviation Training Enterprises at Santa Monica Airport in California. The instructor was a fellow named Doug Myers, and in the course of a day, he managed to clean away all forms of accumulated myth and misinformation that legions of lesser teachers had left in me — aviation's version of waxy yellow buildup. But more than that, Myers taught instrument flying the way it was meant to be taught, as if everyone could and should fly IFR, and moreover, that they should learn to fly with real-life knowledge of the way things actually work. It was only natural that I turned to him for counsel when I started writing, and throughout the project, his advice has been a framework around which ideas have been formed.

Cessna Aircraft Company provided generous opportunities to photograph instrument panels, as did various airplane owners in Hendersonville, North Carolina (y'all land at their airport, hear?).

Carl Becker, Dick Holm, and Ben Phillips of Sperry Univac helped with my questions about ATC systems. Ian Wolff of the FAA helped out there, too.

John Winters of Aerosonic answered some of my queries on instruments, as did Floyd Piper of Edo-Aire.

Jim Terpstra of Jeppesen, although not directly involved, has had an effect on everyone in this business and caused us to worry why we can't remember as much as he does about the rules. He's IFR's version of Enlightenment, if there is such a thing.

Since I certainly never invented any of the rules, ideas, or methods that are contained in this book, the only claim I can really make to any form of originality is in the expression of the thoughts; so I feel a debt to the uncountable heroes in airplanes and on the ground who dreamed up IFR in the first place, and to the ones who followed them and made the whole thing work as well as it does.

GEORGE C. LARSON

INTRODUCTION

It may be hard to believe, but there was a time when instrument flying was a pretty scary business. The entire undertaking was, by our present standard, steeped in uncertainty and unpredictability, and pilots who regularly flew "on the gauges" were viewed by their fellow airmen as the kings of the sky. Airline passengers who scheduled themselves onto flights during inclement weather truly had a taste for adventure.

Before World War II, radio navigation was embarrassingly elementary. From our sophisticated vantage point four decades hence, both ground transmitters and airborne receivers of that age appear barely out of the crystal-set stage. Equipment failure was frequent and sudden, and even when working in top form was plagued with puzzling anomalies. Air-to-ground communication was awful and predictably became worse when it was needed most — during foul weather. Earphones in airplanes often seemed to have an independent supply of static that could be released during reception of a crucial transmission. Weather analysis and forecasting was pretty much limited to what was actually happening and what, statistically, might be expected. The airplane itself — particularly the engine — was still earning its reputation for reliability.

While the rudiments of instrument flying were well understood by a small band of dedicated professionals in the 1930s, it took an unusually intense student to wring the hard-won truths of Instrument Flying Rules (IFR) from the gnarled veterans who guarded their sooth as though it were the family treasure. Bits of information could be picked out of the endless lore by a novitiate through the medium of perfect attendance at hangar-flying sessions, and familiarity with the instruments themselves could be tediously built with hours of cockpit time and weary journeys to nowhere in the primitive Link trainers. But the way one learned the basics was at the knees of an experienced instrument pilot. It might take years of such observing and listening before a student could begin to feel proficient. One could become a professional instrument pilot only if one had the required patience and humility.

There was neither the time nor the need at the start of World War II to indoctrinate the flood of new military pilots in the black arts of instrument flying. For the most part, the flying these young men and women would be doing required instead that they be able to *see* — either the ground or another airplane. Instrument training was necessary and desirable primarily for getting safely out of cloud in the event that they inadvertently should enter one. As a result, for most of the highly skilled combat pilots who returned to peaceful flying, to apply the techniques of instrument flight in actual weather was to operate in a condition of imminent emergency.

By 1944, the number of combat veterans who were returning to the States and trying to adjust to the ways of cross-country flying in any and all weather was growing — but so was the fatality rate. Clearly, something was terribly wrong. As it happened, one of the cadre of thoroughly experienced instrument pilots was the commanding officer of the training facility at Columbus, Mississippi. In one of the more praiseworthy decisions of the time, the Army Air Force Training Command appointed Colonel Joseph Duckworth to establish an instrument flight training school for experienced pilots.

For several years, Duckworth had been experimenting with the best way to indoctrinate his students in the techniques of instrument flying, and his success was measurable by the extremely low accident rate of his graduates. His central theme was that a flight should be performed in the same fashion whether the sky was clear or cloudy, by night or by day. He downplayed the daring and zeal but emphasized thoughtfulness, planning, precision, and pilot responsibility. He taught pilots to fly to predetermined tolerances, which in his case were plus or minus twenty feet of desired altitude and plus or minus three degrees of the appropriate heading. Arrival should be no more than three minutes off its estimate on a two-hour flight. Of course, a pilot must have a thorough understanding of the airplane and its systems as well as the weather conditions along the route of flight and be able to maneuver his airplane smoothly into the most effective attitude to achieve a given result.

This, then, was the beginning of attitude flying, and while it stressed positioning the airplane primarily by instruments and by appropriate use of power and controls, it also aimed to develop confidence through practice and knowledge, so that instrument conditions were no longer cause for apprehension. Duckworth's assignment was to expand the attitude-flying concepts and to standardize instrument flying practices and techniques for the services. He quickly attracted a band of dedicated men, all of whom had had teaching experience. They became the faculty of the Army Air Force Instructor School (instrument pilot) at Bryan, Texas. The school opened in 1944 and operated continuously for over two years. It was without doubt one of the finest military schools of the day. It accepted only volunteers.

In spite of its being part of the military system, discipline, although present, was restrained. The atmosphere was more academic than regimented, and informal hangar-flying discussion and debates were an integral part of the curriculum. Many of the concepts were strange and new to many of the pilots and often at odds with what they had been taught. In some cases, it took as much time to unteach nonstandard techniques as it did to imbue the students with the new truth. But gradually, the effectiveness of attitude flying became recognized; Bryan graduates carried the gospel to the other services and, ultimately, to the more benighted airlines and civilian flying schools throughout the world.

Commercial carriers in the postwar years quickly established a record of training and pilot professionalism that allowed them to operate with a level of safety in instrument flying that was almost the equal of visual, fair-weather operations. General aviation, though, worked to a lower order of constancy and reliability in bad weather. Militating against proficiency by most general aviation pilots were several factors, the most prominent being the abysmal training that accompanied the GI Bill learn-to-fly program in the immediate postwar years. Most airport operators and newly established flying schools were just unprepared to deal with a serious, professional instrument flight regimen. Instructors were hard to find and harder to hold, and examiners' were sometimes hard to convince that an ordinary businessman-pilot or serious-minded pleasure flier should have a license to fly in IFR weather.

Another major hurdle in general aviation IFR proficiency was the scarcity of suitable avionics and the astronomical prices for what was available. It seemed that the several manufacturers of low-price gear were not interested in developing good, high-quality avionics that could perform in an IFR environment, and manufacturers of the equipment the airlines were using were unwilling to sell and service their avionics aggressively to general aviation pilots. It was not unusual in those days to hear it said by manufacturers for the airlines and the military that it was just as well that general aviation pilots did not attempt to fly in heavy weather.

And then there was the intimidating effect of an air traffic control (ATC) system that was designed to meet the needs of commercial airlines and whose users and operators knew it. General aviation was often looked on as an uninvited guest and treated accordingly. While controllers knew well that every aircraft in the system had to be dealt with on an equal basis, they also knew how to demean with a subtlety born of practice any performance that was less than professional. As the system grew and requirements for more equipment increased, general aviation pilots stewed and whined, complaining that they were being barred from the airspace. In reality, the system was growing primarily to accommodate the ever-increasing number of determined general aviation pilots who were finding their way into IFR flying in spite of the odds against them.

By the early 1950s, a breakthrough in avionics manufacture brought high-quality, lower-cost gear within the reach of most general aviation pilots. Everything was solid state and high technology, and the competition became fierce. An avalanche of publicity sold the blessings of instrument flying capability, and pilots responded by regarding equipment redundancy as not only possible but obligatory. With more equipment came more confidence, and more confidence prompted greater use of the system. By the mid-1960s, general aviation use of the ATC system as measured by traffic movements at airports far exceeded that of the airlines. Gradually, a web of mutual dependency between the air traffic controllers, the avionics and airframe manufacturers, and general aviation pilots developed. We are deeply involved in its ramifications today.

The overriding truth of instrument flying these days is that most of it is done by serious, well-intentioned general aviation pilots flying aircraft that are superbly equipped — perhaps even overequipped — and whose overall cost almost *requires* that the airplane be used for all-weather transport. The exorbitant costs of the modern airplane, fully equipped, sometimes conspire to tilt a pilot's better judgment, to urge him to go when he should not. Unfortunately, it is all too easy for the pilot to imagine that with all that airplane and all that equipment, anything is possible. It *is* — if the pilot has a capability that matches his aircraft.

Airframe and equipment manufacturers are today producing superb machines and near miraculous communication and navigation devices. Together, they provide the potential for as nearly perfect an all-weather air transportation vehicle as can be expected. Controllers recognize and fully appreciate the fact that the more general aviation aircraft that fly IFR, the more they are needed and the more the system will grow, primarily with general aviation airplanes as the principal users.

If there is a weakness in the scenario of steady, happy growth, it is the often less than satisfactory IFR performance of some general aviation pilots. And it must be acknowledged that far too many IFR-rated pilots defer using their ratings by straining to stay visual in marginal weather. They like to think they are saving the rating for use in an emergency, and alas, it often turns out that when they do elect to use it, an emergency it is. For these pilots, the rating is like a safety net that is rent with holes. Only frequent use and familiarity with the system can make the rating a passport to safe and better flying.

In the same breath with the above, it must be said that there are a great many excellent instrument-qualified pilots in the ranks of general aviation. Some with whom I have flown are, in my view, the equals of the most experienced military or commercial pilots. Those who take their flying seriously live by the rules and know their stuff. It's just that there should be more pilots like them.

It is also unquestionably the case that every pilot today who has the intention to use a general aviation airplane to practical effect must have an instrument rating and must be able to fly with a thorough understanding of the requirements. To do less is an affront to the system and a threat to the well-being of the other pilots with whom he shares the sky.

Abetting any well-intentioned and aspiring instrument pilot today is the fact that both the modern airplane and the available equipment are, if steeply priced, both highly reliable and marvelously effective. There are some excellent ground-school courses being offered at full-time schools and at modern, well-equipped fixed base operations. And of course we must with no apology recommend this book as one of the more complete and best organized of the literature of instrument flying.

No sooner will you get into the procedures of flying on instruments than you will doubtless be stunned by its apparent complexity and detail. While it is true that the more you can remember of the rules and regulations, the better you should also know that there is a grand and sweeping logic to IFR flying that suggests that if you can't recall every flyspeck of a rule, common sense will most likely prompt you to do the right thing. That's the way most IFR flying is accomplished by most people, because few pilots could recite every Federal Aviation Regulation dealing with every conceivable condition in which they might find themselves. Having made that disarming statement, I should also point out that the better your understanding of both the letter and the intent of the regulations, the better and safer your flying is likely to be.

One doesn't have to do much flying in the system before the reasons for this become apparent. After listening to an hour or so of conversation between controllers and experienced pilots, you'll detect the gentle tone of respect that each has for the other. You will see how neatly the two roles blend and how careful each of the partners is to avoid impinging upon the responsibility of the other. Even when there is a minor breakdown in the politeness and understanding that is at the heart of good pilot-controller communications, there is more often than not an on-the-air settlement of the situation that instantly clears up matters and gets everything back to normal.

I can recall an airline trip I made recently on one of those wide-bodies in which the air-to-ground communication can be monitored on the passenger headset. It was a red-eye flight leaving San Francisco at nine in the evening and arriving at New York's JFK in the early morning hours. I began listening as we were letting down over Pennsylvania. The weather was heavy, and I soon detected that the controller was busy and a trifle testy, perhaps because most of the pilots he was working had been flying all night and were a little less than alert. Our pilot twice was laggard in acknowledging vectors, and the third time the controller snapped, "C'mon there, 647 Heavy, let's get with it." Without an instant's hesitation or a trace of abjectness, our pilot replied, "Sorry, sir," and the system smoothed out without a ripple.

Newcomers to the system are often surprised at how frequently the word "sir" is used in normal communication. It's not a stiff, military usage at all but a genuine acknowledgment of respect for the professional conduct being exercised by each party. And it can be used by controllers to remind an indecisive pilot that it is up to him to make a decision and time he did so, as in "What are your intentions, sir?" A very useful word in instrument flight, "sir." It's an implied salute between equals suggesting that if you do your job, he will do his. And perhaps the most telling contribution of Joe Duckworth to the art of instrument flying is that he believed that doing the job right was really quite within the talents of most of us, that precise instrument flying, when well planned, smoothly flown, and thoroughly controlled, was not only hugely satisfying but also great fun.

ROBERT B. PARKE

How to Use This Book

Instrument flying is largely a matter of rules and procedures, not a "style" of flying, just as flying very large airplanes like airliners is mostly a matter of learning to manage their systems. This book was never meant to substitute for the core of knowledge that forms the foundation for an instrument rating, so in that sense, it's not so much a curriculum as it is a personal view of IFR.

It's also been my experience that after a couple of thousand hours logged, most pilots don't have much need for the advice of somebody else. That's not egotism but the realization that all the advice they've ever gotten hasn't meant a fraction of what they've experienced themselves; some of life's lessons have to hit you over the head. So this book is aimed pretty directly at those pilots and other people who wonder what instrument flying is like and can't really find a single source that describes it *all* to them.

The book is also centered around the performance and behavior of light airplanes, the kind we are most likely to learn in. I suppose that someone, somewhere has actually gotten his or her rating in a turbofan-powered cosmic racer, but most of us ordinary mortals undergo our initiations in more prosaic airplanes, with their attendant realities and limitations, the most important of which is a rather mundane service ceiling. There are such things as mountains and weather systems that we cannot surmount, our personal virtues notwithstanding.

Following each chapter that deals with rules, there is a section called "IFR Notebook," and it is proof of the fact that this book is a personal work and not without its tendencies to cater to my own biases. It bothers me — always has, always will — that to find what you need to know to fly IFR (and even some of the important facts that are merely nice to know), you must undertake a literary scavenger hunt the likes of which no other profession imposes on its members. The rules and structure of IFR are diffuse and secret, and the sources of fact are elusive and hidden. It seems to me that if you buy a book like this, I owe it to you to bring together between its two covers at least those articles of IFR convention that gave me the merriest chase.

Now, there is a certain caution that goes with the luxury of finding lots of facts like these in a book like this one. Every so often, the rules are revised, and unless you are the government or Jeppesen, there is no way to keep up with the tidal processes of change. I've tried, insofar as possible, to list the facts that seem most likely to endure. It will be your responsibility, however, to ensure that you study the most recent regulations and procedures. Use what I have included as an outline for those areas you should understand — you'll never remember all of them all the time (and anyone who says he can deserves all your suspicion) — and simply be guided by the Notebook, not bounded by it.

It is my unashamed purpose to have you leave this book encouraged to try instrument flying, so I can fairly be accused of lack of balance if you're feeling doctrinaire about writers' responsibilities. This book unabashedly promotes instrument flying as something in which most pilots should be trained and capable. It is not meant to represent a training course complete and self-contained, though. It may even disagree with some current training philosophies in some minor respects. But I care less whether you agree with everything that's in here than I do that what's in here makes you think about some aspect of flying that you'd never thought out before. And maybe, just maybe, some of that two-thousand-hour crowd will find something worthwhile in that as well.

To be quite honest, writing this book was no fun. I found it most difficult to dredge up my own early memories of those aspects of obtaining an instrument rating that gave me the worst fits; yet I had to concede that those are the areas that are most important when you're in the early stages of learning. There is so much unsaid, so little assurance that others have beheld the same fears you confront right now, so much in the whole system about which they tell you, "you'll pick it up as you go along." Some of what's in here is the unsaid, the un-written, and the unmentionable. I know that right now, if you're staring at the front end of an instrument rating training course, the obstacles of doubt and question, ignorance and fear, seem insurmountable. Everybody else who's been there has felt the same way, and I suppose we can accurately be accused of perpetuating that situation as a form of fraternal initiation rite. It was time that ended anyway.

GEORGE C. LARSON

CHAPTER I

Flying Blind

They used to call it blind flying. Look, ma, no eyes. The term lacked accuracy, but it more than made up for that in titillation, and the newspapers loved it.

The truth is that if James H. Doolittle had literally been *blind* on September 24, 1929, when he executed the first landing using instruments alone for reference, he'd have been a goner. It is only one small example of the apparent contradictions in IFR flying that to accomplish an instrument flight successfully, you must ignore every one of your senses *except* your eyesight.

Instrument flying is not visual in the traditional sense of that word because it makes no use whatever of any sighted reference outside the airplane. In fact, one of the most difficult things for a new pilot to avoid when flying on instruments is the tendency to look outside the cockpit. It's disorienting, and unless you happen to be at one of those rare moments when it's allowed — at decision height on an ILS approach, say — you just don't do that.

Instrument flying substitutes for the usual real-world visual reference — the horizon — several standard sensors and displays, all of them arrayed in a generally agreed upon order on the panel immediately in front of the pilot; the secondary stuff is shuffled off to the sides. Although one of the instrument displays is, in fact, a little model of that external horizon, it is not, as you'll see, the only thing the pilot needs to watch in order to fly on instruments with competence.

Most of the standard training for an instrument rating is designed to make you comfortable in this alien, artificial environment. The training begins by showing you how the instruments work and proceeds by instilling within you an abiding faith that they are far more trustworthy in cloud than are any of the sensations of acceleration that your body may pick up through the nerve endings in the seat of your pants or through the disparate signals from the organs of balance in your ear.

You will hear some instrument-rated pilots tell you that this rating is the most difficult of all to obtain, and then these same folks will turn around and tell you that once you've earned it, an instrument rating makes flying easier than ever. The funny thing is, both statements are true.

What makes the training difficult is the learning of a whole new set of reflexes and responses. You are expected to adapt to a family of displays that, taken together, give you a very accurate picture of exactly what the airplane is doing. When you fly visually, you watch airspeed, manage the engine, and maneuver by looking outside. On instruments, all of this becomes more precise. Instead of turning so you're "pointing the nose at that mountain," you will be expected to take up a very precise heading and to hold that within a degree or so. If you had to rely on outside references alone to do that, chances are you would fly the airplane with much less precision than you would on instruments. Instruction for a private license includes some time under the hood, and you may occasionally check your directional gyro while you're flying VFR, but IFR training will make you realize for the first time how the gyros in that panel multiply the sensitivity of your eyes manyfold. The whole idea behind a standard IFR instrument panel is exactly that: to provide your eyes with displays on a scale to which you can comfortably respond. It would be very difficult, for example, to produce on visual references and senses alone the correct angle of descent to give you a rate of 500 feet per minute downward. You weren't born with a sensory organ that can provide that sort of information, which is exactly why there's a vertical speed indicator (VSI) in the panel. Similarly, when you can't see anything outside but the gray gut of a cloud, you are mightily pressed to state with any degree of assurance, "This is a standard rate (180 degrees of heading change per minute) turn," even though your body may be telling you you're in *some* kind of turn.

The scale around which all of these instruments is designed is based on the way you'll actually be flying. In the case of the vertical speed indicator, for example, the scale on most lightplanes reads from zero to 2,000 feet per minute, climb or dive. You don't dive airplanes of the type we use at rates that ever exceed 2,000 fpm, and most of their engines won't give you a 2,000-fpm climb. On the other end, you'll find that the scale is divided into hundreds of feet, the smallest division on the VSI. The reason for that is simple: rare are the situations where you have to provide climbs or descents in, say, *tens* of feet per minute, and if you should, an approximation will suffice. Thank your predecessors for figuring all this out, because back there sometime, there must have been some poor guinea pig who tested a vertical speed indicator marked in tens of feet per minute. He must have overcontrolled his airplane something awful. ("Hey, Gerard, tear that damn fool thing out of there and get me one marked in hundreds." That was progress.)

So this new set of reflexes you must develop within a rather alien environment is really only an adaptation to new cues that come to you from instruments that are time-tested. They provide you with responses to what the airplane is doing, and these responses, furthermore, are displayed on a scale that is far more sensitive than the cues you learn while flying visually but not so sensitive that you'll be overreacting. An instrument panel is a compromise in this respect, and although someone may improve on it someday (people are always trying), the present model seems to get the job done.

It doesn't really matter what category of aircraft you are flying — an airplane from Stearman to SST, a helicopter, or a blimp — because the standard instruments make it possible for you to manage the whole show by telling you exactly what's happening in all the important axes of motion. You'll always know exactly how fast you're moving through the air, where the wingtips are in relation to the horizon, where the nose is pointed with respect to magnetic north, how high you are, and whether or not you're flying straight (no yaw). Add navigation instruments to that and you'll even know where you are with respect to a point on the ground.

In instrument flying, you first learn to control the airplane using the instruments, and later, once you're good enough to do that by reflex, you add the overlying problem of navigating to a specific point along a planned route. And believe it or not, that's really all there is to it. Veteran instrument pilots to whom the manipulative part of the exercise is simply second nature see flying IFR as a test of memory (the rules), handwriting (copying clearances as fast as controllers read them), and ingenuity (getting the system to work for you instead of the other way around). Some of them even say an instrument rating is nothing more than a "tool" that increases your odds of arrival once you depart. Admirable. Calvinistic, even. But there are moments when IFR flying can be satisfying for the sheer achievement of it; another word for it is "fun." So accept the utilitarian view if you like, but you'll have to forgive the rest of us if we insist upon enjoying ourselves. We have no plans to force the others to see it our way, and anyway, it's not very popular these days to admit that you're actually having a good time.

I happen to like IFR flying, and one of the purposes of this book is to persuade you to like it too. When you lack an instrument rating, the whole thing looms large before you. As a pilot, you feel a certain delinquency. And then there's that tacit ethic, kind of silly, really, but it says that a pilot is the sum of his or her ratings. Nobody expresses it quite that way. They like the negative argument better; that is, you're not a *real* pilot until you've flown

- taildraggers
- on instruments
- multiengine
- jets

and you know all the rest. If you feel you must, get an instrument rating for that reason and get it out of the way; maybe it'll liberate you from that sort of nonsense from now on.

An instrument rating is really worth earning because it represents mastery of a set of professional rules and disciplines. Instrument flying is a task that provides simplicity through complexity (and at awkward moments, complexity through simplicity), and it will probably make you see yourself in a different way when you've emerged from the experience. In other words, you'll grow a little. Oh, it's no encounter group or therapy or anything like that, but it has a certain quality to it that sets it apart from the other ratings you can collect. Ask any instrument-rated pilots and chances are they'll tell you that the two experiences that stand out in their memories are their first solo as a student and their first instrument flight when they flew it down to minimums and looked up to see the runway. The first time you do it in your training, you will turn and look at your instructor. It's an involuntary reflex, and everyone does it. Not a word will pass between you, just a smile and an unspoken "Oh, so *that's* what it's all about."

There are quite a few instructors who'll teach you to fly instruments and then turn around and say something like, "Now don't actually *use* your rating for at least a year. Fly VFR and file, do all the approaches visually, and whatever you do, don't fly to legal minimums. Add a couple hundred feet or so." They're entitled to their opinions, but that's all such advice is — an opinion. Here's my opinion: if you fly around VFR-only for the year following the date of your rating, you will not be proficient to fly a real instrument approach. Maybe that makes sense to you, but it doesn't to me. Either you are rated or you aren't. There is nothing dangerous about the final 200 feet of an ILS approach that will bite you just because your rating is new, and many pilots are really at their best right after the check ride. If you are so bad that you can't even recognize that you are flying the approach badly and should execute a missed approach, then go seek further instruction. I can't imagine why you would depart from any instructional course in instrument flying and feel that it was anything but a colossal swindle if the school tells you *not* to use the rating. End of opinion.

You'll find that most pilots, lunatics excepted, are self-governing on the matter of currency and confidence. All of us have flown with someone who is obviously shaky, and in most cases, that individual will quietly seek out some training and solve his or her problem. There are regulations that govern currency, and then there is fear. What you will deal with most of the time is the regulatory pressure of fear, or call it lack of confidence. When it happens to you, you'll know it, and it will urge you to the right corrective actions. If you have some kind of death wish, you probably won't be reading this anyway.

Although you may be so impressed with the difficulty of flying IFR that you think it's out of your reach, be encouraged that about a third of all active pilots are instrument rated. Surely if that many people were able to do it before you, you can do it now.

So my advice would be to forget the "blind flying" aura that makes all of this sound like a stunt. You are not training to become a tightrope walker. You will learn some pretty exciting skills, among them an entirely new language (jargon actually), some rules that begin to make basic good sense once you work with them, and a new and more pleasant way of regarding the airplanes you fly.

One way of illuminating the experience of flying on instruments is to look at a machine called a flight director. The flight director was invented to help pilots of large, high-performance aircraft to fly smoothly. Flight directors are glorified artificial horizons that superimpose a little display — usually paddles or pointers — telling the pilot what to do with the wings and the nose: bank left, descend, climb, or whatever. In other words, if a rated instrument pilot

A typical flight director; this one happens to be a Bendix director/horizon indicator that forms part of an automatic flight control system. The two darkened wire struts that can be seen on both sides of the instrument's face connect to the two cueing paddles. The airplane symbol is the slightly darker wedge-shaped device sitting atop a vertical wire strut. It's your job as the pilot to manipulate the controls so that the airplane symbol is always nestled in the position with respect to the cues shown here (this is what it looks like in straight-and-level flight). In flight director terminology, the cues can "command" climbs, descents, or banks along with combinations of those maneuvers. Navigation information such as heading and course deviation is referred to as "raw data," and the computer that drives the flight director cues is programmed to combine all of the data into a single command and to do it very smoothly. The pilot still selects the correct electronic course guidance and magnetic headings. It's the flight director's job to indicate how to fly the airplane so that its course matches the one you've selected.

were to operate the flight director for you, even if you were a pilot with no ratings at all, not even a private license, you could successfully land an airplane
simply by following the command displays on the flight director. An autopilot
is nothing more than the flight director computer connected to some kind of
servomechanism that operates the control surfaces for you. Now they don't
even need your hands; you just sit there.

You have to admit, though, that a machine that can fly an airplane is quite an
accomplishment. That it can probably fly better than you is humbling, but no
less true. Now for the really humorous part: you can't get your instrument rating by flying with the autopilot or flight director turned on. That would be too
easy. Even though flight directors are routinely being installed in light single-
engine airplanes these days, the traditional method of hand flying still prevails
when you go out to take a check ride with the inspector.

The government is not trying to be difficult; the truth of the matter is, no
system aboard an airplane has ever been intended to *replace* the pilot, only
make the work load easier. Flight directors (and autopilots) were invented to
do exactly that — ease the work load — and they are not a substitute for basic
flying skills. Most pilots actually look forward to the fun of flying an approach
after the boredom of the en route portion of an instrument flight; coupling the
autopilot deprives you of one of the most satisfying portions of the trip — keeping everything centered all the way to touchdown, and all by yourself.

What I find intriguing about flight directors is that they provide a nice philo-
sophical hook around which to organize your thinking about instrument
flying. After all, a flight director accomplishes the central piloting task of instrument flying. It assimilates all the sensory data from the instruments (the
attitude of the airplane) and the navigation displays (your position over the
ground) and comes up with a firm decision about what to do next. If all instrument pilots flew like flight directors, the IFR world would be a perfect place.
What is even more amazing, when you think about it, is that *any* flight director can fly *any* aircraft. All you have to do is tune it to the' airframe so it
knows how much the whole affair will move for a given control-surface displacement. (That's what they talk about when you hear them saying they've
got to "adjust the gain" on an autopilot. Genuine hangar talk.)

This little dickens of a flight director is really pretty talented, and its particular skill, the one we want to emulate, is its ability *to integrate all the sensory
data and make a smooth correction to obtain a desired result*. I don't think I
know of a better definition of a good instrument pilot. If you can absorb all that
stuff, make sense of it, then ever so silky smooth, turn, descend, or climb to
get where you want to be, you have done it all. The more you look into flight
directors, the more you've got to admire them.

You see, the only thing that makes instrument flying tricky at all is that without an outside visual reference, the pilot is deprived of the only source of infor-

mation about rate of change of position of the airplane. When you can see the ground moving beneath you, the horizon in front of you, and you keep your airspeed within its limits, flying an airplane is no great feat of coordination. What you really appreciate about a visual reference, once a cloud or a training hood deprives you of it, is the feeling of motion it gives you. On instruments, you feel occasional tingles of acceleration, which you're supposed to ignore anyway, but aside from that, you could be standing still. That's what gives instrument flying such a "remote control" feeling. As it happens, it also enables an absolutely stationary simulation device called a ground trainer nearly to duplicate the experience of an instrument flight without flying. Add even partial motion to a trainer and the "flight" in a *simulator* can be very realistic indeed.

Since the airplane behaves the same, visually or on instruments, the instrument pilot has to find a way to substitute for the missing sensation of motion. After all, in order to make the flight, you have to be aware of where you are and how fast you're moving. You can't land the airplane at your en route cruising speed, and the airplane will have to be aligned with the runway. In a high-performance airplane, you must be aware when a fairly sharp turn is coming up so you don't overshoot badly. So it's an important attribute for an instrument pilot to be able to create a mental picture of the airplane's position and its rate of change of position. Doing push-ups every morning won't help you to develop this ability, and pure strain just won't produce it. It will simply happen. Without thinking about it, you'll find yourself creating a picture, sometimes from outside the airplane looking down, sometimes from inside the airplane, imagining where the airport lies or where other traffic may be. You'll shift between various views of yourself as you find it necessary to solve the problems that confront you.

You will find with experience that a cockpit in instrument conditions, with its lack of real-world stimuli, makes it fairly difficult to create a picture of your position at any given moment; but what is more difficult to create is a *dynamic* picture of the situation.

Airplanes have one problem: they never stand still. Even in a holding pattern, where you'd think there'd be plenty of time to breathe easily — after all, you're not really going anywhere — things happen fast enough to challenge the ability of any pilot who hasn't practiced holding. You change heading once a minute, the wind may be affecting your course over the ground, and you may be performing some furious knob twisting if you have only one radio.

Here's where we come to the connection with the flight director: learning to fly on instruments is largely a matter of acquiring the habit of picturing a constantly changing situation, a situation in which the rates of change may seem to alter dramatically; from this dynamic picture, you must integrate two ideas: where you are now and where you ought to be. From that comparison, you decide upon a maneuver to make the two match up and then execute the maneuver. The flight director gives you command cues to follow, and if your wingtips match the cues, you have solved the problem. When you, the pilot, fly like a flight director (even though you don't have one or it's shut off), you perform exactly the same function from moment to moment.

It's easy to understand the situation when your real wingtips match those of the cue (either the flight director's or yours); no maneuver is required. Status quo. Beautiful. If only it were all this easy. When you are all settled down en route, in trim and on heading, you can turn the controls over to a completely nonrated pilot and say, "Here, fly instruments," and the nonrated pilot can do exactly that by doing nothing. I almost hate to tell you this, but for 99 percent of the normal instrument flight, that's exactly what you do with the airplane: nothing. And when you do something, it will almost never be very abrupt and certainly not thrilling. The actual manipulation of the airplane's controls on an instrument flight would put a Tasmanian devil to sleep. Actually boring. The reason for it is simple: instrument procedures are designed to be very gradual in terms of rates of change of heading, altitude, and airspeed. Since you're deprived of the ability to see how everything's changing and you must construct a constantly changing picture of it as a substitute, the folks who design the whole system have tried to make it as easy on you as they could. Instead of a roller coaster, you'll be riding something more like the stretch of track between Atchison and Topeka.

There are times when things do change, though, and it's those moments when the *rates* of change are altering most quickly that are the most challenging to fly and to visualize. As I already mentioned, flight directors were invented when airplanes got so fast or so big (or both) that the rate of change was just too much for the pilot, and he or she needed some help to put it all together.

If we, as human pilots, could handle rates of change as a flight director does, we'd really be in good shape. First let's examine the dynamic situation in an instrument flight to find out why the rate of change creates such a problem.

Let's say we're en route between two cities with a pair of VORTACs at their airports. Let's also say the airway connecting the two is a straight shot, with allowance for magnetic compass adjustments. We depart the first city and climb onto course until the course deviation indicator needle on the omni display is centered and the correct heading has been selected. A while later, we decide to check on the fuel consumption at this power setting, and when we turn our attention away from the piloting task for a moment or two, we fail to notice that we've changed heading slightly. After flying at the wrong heading for a minute or two, we recheck the CDI needle and see that we're slightly off course. We could correct this error in any way we please, so let's examine the situation. Only slightly off heading for two minutes means we're probably pretty close to the course, probably off by something less than a quarter of a mile (we could figure it out, but that's not the point right now) and maybe even just a few hundred feet.

So we know we should make a turn back onto the course, but we should also think about how fast we should correct. A 45-degree bank would certainly do the job. We'd be hauled over into a turn quickly, make the correction quickly, and — whoa! That was *too* quickly. We just learned a first lesson: rapid, hard corrections are not appropriate right now. A little 5-degree bank to turn us gradually over to a new course that will slowly bring us back onto the proper centerline is a much more pleasant and efficient way to solve the problem of the mismatch between where we are and where we should be.

You can see from this simple example the problem that confronted the designers of the first flight directors, and autopilots. Suppose they'd elected to wire the needle from the CDI to a switch. Every time the CDI needle deviated even slightly from center, the switch would order an abrupt turn. We've already seen that a system like that just wouldn't work. Flight directors and autopilots had to be based upon a model of the actual dynamic situation in order to work successfully. They not only had to command the correct response from the pilot or the controls, but the response had to be matched to the situation so that its *rate* was appropriate. If you take apart an actual course-correction maneuver, after all, it's really pretty complex: decide which way you're off course, roll toward the course, roll out on new correction heading, anticipate approach of course centerline, roll to intercept, roll out on centerline.

The fundamental problem for us in locating our airplane in space has to do with the system of geometry we work with. All the present generation of navaids, with the exception of area navigation computers, work on polar coordinates, which is just a fancy way of saying they are based on the degrees of a circle. Automatic direction finding, omnidirectional radio gear, and instrument landing systems with their localizers and glide slopes all read out to the pilot in *degrees* of displacement from centerline. It doesn't take much talent at visualization to realize that two degrees may represent an actual linear distance of several feet when you're close to the hub and several miles when you're far away from it. It's important to get that absolutely straight in your mind at the outset, because it has a lot to do with why things appear to change so fast when you're close to a station and why they seem to change so slowly when you're far away from it. Part of the challenge of learning to fly on instruments is learning to interpret what you see in your navigation indicators in light of your picture of where you are with respect to the station.

A flight director is able to do this, and it took a genius of a man to construct a mathematical model around which he could design a computer to do exactly what you're supposed to do with your hands after you read this book and get your instrument rating. Any off-course situation represents a quantitative error. Whether it's a heading error or a linear distance, ten feet left or right of a localizer course, say, it can be measured by an electronic signal in a directly proportional and quantitative way. The point is to reduce that error in a gradual, smooth fashion, so that nobody notices. It turns out that there is a curve in mathematics that fits that specification exactly. The curve is called an *asymptote*. Mathematicians know that asymptotes would never actually touch the course line but only approach it very closely. (Asymptotes do touch the course line at infinity, but there is no published approach for infinity anyway.) It's much easier to design a machine if you can strike a mathematical model to copy, and it happens that asymptotes are the curves most similar to our ideal course-correction maneuver.

By combining electronic signals from the navaid receiver, from the compass heading, and from the amount of bank (or roll) on the wings, a flight director can call for course corrections that are always perfect, no matter what its distance from the station may be. In other words, it is capable of handling multiple situations with entirely different rates of change. The designers have added refinements to flight directors to aid in this "smoothing process," as nav system designers call it, so that on some of the most sophisticated jet aircraft, you can fly across a station and get a full-scale off-course deflection of the needle on a nav-receiver indicator, the kind that might urge a brand-new IFR student to haul the airplane over into a tight turn. But the flight director is a mechanized veteran, and it will ever so smoothly command a gradual roll to just the right heading to intercept the new course.

Even though the kind of lightplane you're flying is not that fancy, the "flight director method" of learning to fly on instruments will enable you to achieve this same smoothness in your own hand-flying. It's more than a way of learning to manipulate the controls, though; the flight director also gives you a valid model for a way of thinking about all the aspects of instrument flying in terms of rate of change. It can help you to communicate in a professional manner. By "professional," I don't mean the use of words like "no joy" and other arcane nonsense. Communicating professionally means getting your message across clearly, the first time, using the minimum transmission time that's necessary to do the job. It also means having a clear idea of what you want and anticipating the options you may be presented in the reply from a controller. If you begin your instrument flying with this way of thinking, or if you are already an experienced instrument pilot and can incorporate even portions of this philosophy into your flying habits, you'll find a great deal of pleasure in each flight under IFR conditions. As if a flight director were aboard, you will be unhurried, sure of yourself, and best of all, a smooth pilot. And since the IFR system is already constructed in a way that eliminates the need for any extreme maneuvering, you'll be tuned precisely to the real world of instrument flying.

CHAPTER II

Learning How: Schools, Teachers, and Books

When you're in the eighth grade, you go to school because you *have* to; it's the law. But nobody is making you learn how to fly (sometimes it seems as if the world would rather you didn't), so most flight training starts from the assumption that you are self-motivated. Nobody's saying that's the way it ought to be, but the fact is that some of the best and worst aspects of aviation training are founded in the seller's-market mentality that currently prevails in the schooling of pilots. It's probably true, therefore, that your success as a student in an instrument course will be largely up to you.

Since you cannot get an instrument rating without a valid pilot certificate — meaning at minimum a private license — you will already have been exposed to some flight training when you begin your instrument flying education. The gap between those hours necessary to achieve a private pilot license and the experience required for an instrument rating used to be filled with the commercial pilot course; it practically guaranteed 200 hours' total time on the day when you started instrument training. If you choose to go without a commercial pilot license, you'll have to find other ways to build experience (which means "hours," really, and *not* "experience") in order to meet the 200-hour total time requirement and 100-hour pilot-in-command time. Just renting a $35-per-hour single for a hundred hours costs $3,500, so this is no minor intermission.

Instrument training falls readily into two distinct "departments": Department F and Department R. The F in IFR stands for Flight, and it will seem the easier of the two only because most of us like to fly. The R in IFR — the Rules — have been described by more than one instrument-rated pilot as the only gristle in the roast. And it is fair to say that the R department will test your motivation more than will the F. For a couple of reasons, most instrument training is scheduled so that you take care of the R first and then go on to the F, with a certain amount of overlap, of course. For one thing, it's part of human nature to get the less enjoyable tasks out of the way before proceeding to the fun part.

In the case of instrument flying, it is also true that you might waste time and money in Flight if you lack fundamental knowledge of how the Rules work before you set out to use them.

As with most things in life, there is a right way and a wrong way to attack the problem of learning instrument flying. The wrong way is, by definition, any way other than the right way. Flying, particularly instrument flying, is not all head learning and not all eye-hand coordination but a combination of both; therefore, learning it properly is a process unlike any schooling ever encountered. For the sake of argument, let's compare it to learning how to weave cloth, another complex skill requiring manual dexterity and the ability to visualize a completed product. If you set out to learn how to weave, you would almost certainly begin by watching someone who knows how. One can only wonder at beginning instrument students who come to the subject without ever having watched either an instrument flight or the air traffic control (ATC) system at work. It is unrealistic to expect to learn to weave well by reading a narrative of how it's done or by listening to someone describe it verbally. It is just as unreasonable to expect yourself to perform well during your IFR education with absolutely no exposure to the instrument system prior to your first lesson.

Chances are that during instruction for a private license, most pilots have visited a control tower cab; almost any wise instructor would fit that (unrequired) visit into the curriculum somewhere so that the student can observe how a tower crew functions and get a chance to see the world as controllers see it. (If you haven't done it, it's something you ought to take care of before proceeding much further. Control towers are listed in the phone book under "U.S. Government, Department of Transportation," and don't worry about the chance that you'll interrupt them in the middle of some *Airport* movie Mayday where they're talking a six-year-old down in a 747. The person who answers the phone probably won't be actively controlling traffic anyway.)

Before delving into an instrument course, you should see how the other half of the system — the air traffic control organization — works in the real world. Unless you happen to live near one of the twenty air-route traffic control centers in the continental United States (or the ones in Alaska or Hawaii) that control en route traffic, your best chance to observe a radar-based traffic control facility will probably be the radar approach control associated with a large airport close to you. It may be smaller and somewhat less sophisticated in some respects than a full-blown center, but it will give you just as accurate an idea of how radar traffic control operates. Many air traffic control towers are now equipped with BRITE radar too, and although towers operate a little differently from approach controls, you'll see how radar presents the traffic information. Since you will almost always be working with controllers in a radar environment in the United States, it's important to know what life is like from their point of view. Until you've physically visited a radar facility, you simply cannot comprehend the task of your partners in this team effort of getting an airplane from one point to another under instrument flying rules. The FAA has a program called Rain Check that allows for regularly scheduled visits to ATC facilities by groups of pilots, and as soon as your name appears on the active pilot tapes at the computer center, you should begin to receive mailed invitations when these tours are held. If not, call and ask about a visit anyway.

During this preparatory period before you begin IFR instruction, make an effort to ride along during some instrument flights, in the right seat if you can, while an experienced IFR pilot operates within the system, and try to participate in some limited way that is acceptable to the pilot. Be considerate while you ride as an observer, though. You may find you have a lot of questions, and you want the answers right now! Try to hold them for those times when the pilot is not terribly busy. It also helps if you can arrange to borrow a headset (or buy one — you really ought to have it anyway) so you can listen to the radio chatter as the flight proceeds. Otherwise, ask politely if the captain wouldn't mind using the cockpit speaker system so you can hear what happens. The reason you want to be polite about it is that most experienced IFR pilots prefer the talking earmuffs to the speakers, which are almost uniformly awful when you want to understand every word clearly. Don't be bashful about seeking opportunities to observe operations in this way, as there is nothing at all unusual about it. In fact, many of the big full-time schools incorporate observation rides into their programs of instruction; two students ride in back while a third flies left seat.

The important thing to learn at this stage is not the particular details of IFR but the pace and rhythm of things. Even before you set out to watch some IFR flying, you should try a short, meditative exercise: whenever you happen to think of doing it, time exactly one minute by watching the sweep second hand go around or by counting to 60 with "Mississippi" between the numbers. You'll discover that one minute can be an astonishingly long period of time. It also turns out that one minute is the briefest unit of timing required in IFR flying: the legs of a holding pattern are one minute in duration. In the excitement of your first attempts at instrument flying, you'll find yourself in a great rush when it's not really necessary. That's when you should think about your armchair minute and its leisurely passage. Take your time in IFR flying. The system is designed to operate at a comfortable pace, and even when you progress to high-performance aircraft, you'll find that awareness of your situation will always provide you with sufficient time to manage. Even when you get behind the airplane, which happens to everyone now and then, the proper reaction is not to double-time every move but quite the opposite: slow down and recover. The system even has ways to allow for mistakes (one is called a missed approach, which more than one pilot has declared almost before the approach phase has even started) so that you can come around for another try. The point here is not to encourage slipshod flying but to emphasize that instrument operations require an inner clock that will only run properly when you achieve a disciplined relaxation about your cockpit routines. If you have a hard time imagining the feeling of "relaxed effort," ask a runner or any athlete. They know that in order to deliver maximum performance under stress, at the first sign of tension, they must relax their bodies by force of willpower. As an aid to you, think about your armchair minute if you find it happening while you're flying.

A home VHF receiver that will tune the aircraft band (108.0 to 136.0 MHz on your dial) allows you to listen to IFR communication if you're close enough to an ATC transmitter servicing a radar facility. Tuning a tower won't teach you that much except on bad-weather days when all the incoming traffic is IFR; on the good days, you'll mostly hear "close up the pattern" and other unenlightening advice. What you want to hear is "marker inbound" and "missed approach" and "What are your intentions?" That's the educational channel. Tune the frequency for the ATC sector where you live and you'll hear at least half the conversations; most aircraft will be audible even if the ground transmitter is out of line-of-sight range. It's the pilots you want to emulate, after all. You'll soon hear enough to know the difference between the people who use the radio well and those who use it badly. If the pros always sound as if they know what they're doing, it's because they *do*. It's your job to sound the same way. Various pilot stores also sell recordings of ATC conversation on records or tapes, but once you've heard them, you'll get nothing new after that. A receiver may cost more, but you'll hear the real thing, and the situations will always be new.

There are several sources of books and publications, beginning with the U.S. government. Advisory Circular 61-8 is the Instrument Rating *Written Test Guide*, and in it you'll find a section entitled "Recommended Study Materials." It contains questions in the style and on the subject matter you can expect to meet in the actual written test. There is only one problem: it lists the possible answers, but it doesn't tell you which answer is correct. For a guide that contains both questions *and* answers, order the *Instrument Pilot Airplane Test* from Aviation Supply and Academics, 6820 Perimeter Road, Seattle, Washington 98108, or call them at (800) 426-8338. The government really can't guide you to any sources of material aside from its own, so their list mentions only government publications. One of these stands out, however, and is a key to any instrument course: Part 1 of the *Airman's Information Manual*; in fact, many of the commercial study courses contain a large percentage of this material merely reprinted. You buy it from the U.S. Government Printing Office. Jeppesen's *J-Aid* also includes *AIM* subjects.

You should also obtain a copy of Advisory Circular 00-2, which is a list of the current publications, revised three times a year, that are available from either the Printing Office (at a price) or from the FAA (free of charge). AC 00-2 comes from the U. S. Department of Transportation, Publications Section, Washington, D.C. 20590, and it's free. It is a compendium of every government publication that has an advisory role in aviation, and that makes it important.

You can buy a subscription to the Federal Aviation Regulations from the government, but there are other sources that may be more convenient. Government issued FARs are published with an unwieldy system of changes. Each chapter now comes as a separate loose packet of pages stapled together rather temporarily with the intention that you keep them in a three-ring binder. The changes to each regulation are published separately and must be introduced by page substitution. Many pilot bookstores sell compendiums of the FARs you'll need to know well: Parts 1, 61, 91, and the National Transportation Safety Board (NTSB) Part 830 regulation for accident reporting. The Jeppesen Company goes one step further and publishes an annual subscription to the FARs that includes a revision service. The Jeppesen format is much more compact — to fit their standard ring binders — and encourages you to carry the Regs with you. For some reason, the Jeppesen format also makes the Regs seem less formidable, possibly because they seem so much more condensed compared to the telephone book you get from the government.

The government's textbook of IFR flying is called the *Instrument Flying Handbook*, Advisory Circular 61-27, available from the Printing Office. It represents government doctrine, and you'll be taking government exams, so . . . The *Instrument Flying Handbook* also is unique in one respect: it bears the *only* FAA advice to be found anywhere on instrument instruction. There is a message there somewhere, and you can't help but wonder why there is no manual of instruction for the teaching of instrument flying. One guess is that the subject simply defies formalizing, that very few instructors would agree on any single method for teaching it. Therefore, the flight test guide and written test guide specify standards of performance. The end, not the means to it, is easier to define.

The *Instrument Flying Handbook* also contains a study outline for the instrument written test, but a much better outline appears in the *Written Test Guide*, AC 61-8. The latter outline contains very specific references to regulations and other publications, so that you can look up the source material for each area in which you will probably be questioned.

At this point, you may also wish to begin your subscription to a chart service so you can observe how one works. Government charts are published by the National Ocean Survey, and they used to be designed primarily for military and other government aviation users. Since the written exam is based on the NOS charts, it may be easier on you to start with them. The NOS has changed its format somewhat in order to make its charts more appealing to a broader segment of aviators. It revises its approach charts once every 56 days (eight weeks), and indicates revisions not yet in force by attaching an effective date.

Every *other* 56 days, 28 days after the main mailing, they mail a supplementary revision, assuming one is needed — and it probably will be. The NOS mails bound booklets rather than loose-leaf pages, and except for the ones covering Florida and Michigan each booklet contains approaches for more than one state. En route charts show airways between terminals rather than approach information. You'll be using these too, and the NOS publishes a kind of index of these and its other materials entitled *Catalog of Aeronautical Charts and Related Publications*. It's free.

Jeppesen charts are published by a private company in Denver, Colorado, and the majority of airline and general aviation users subscribe to the "Jepp" system as opposed to NOS. The equivalent service is slightly less expensive coming from Jeppesen as opposed to the NOS, but there are also several differences in format of which you should be aware. Jeppesen charts are designed to fit a sidebound loose-leaf book, and the revisions arrive in an envelope with a control sheet so that you can check off the mailing to ensure you've received every chart you'll need according to your subscription. The revisions arrive once a week and are inserted simply by replacing pages. A subscription to a particular area should include en route charts, which contain terminal communication information as well as the pertinent airways. Make sure your Jepp service code includes the en routes. (This makes them a very convenient adjunct to VFR flying too, which is worth consideration.)

Do not abandon sectional charts when you begin flying IFR. Sectionals are necessary for visual information, since no IFR chart service provides that. The sectionals are particularly valuable for their detailed terrain information, and you'll see why that's important later in this book.

Various agencies, both government and private, provide airport directories, which you'll probably want to add to your library later, but with the list of publications outlined above, you'll be able to read enough to gain a clear idea of instrument flying and its associated systems before beginning formalized instruction.

It is acceptable, by the way, to use a home study method in order to gain the knowledge necessary to pass the written test. The rules say you must have "logged" home study, but they don't say what "logged" means. Check with the FAA testing center or GADO (General Aviation District Office) near you to get their definition, and then follow their advice.

Prior to the start of flight training, there is a great deal you can do to ensure that you won't waste any time learning to *fly* during what is supposed to be an *instrument* course. Begin by deciding upon the type of airplane in which you will take your training, then become so familiar with its systems and its dynamics in flight that these matters are almost second nature to you. Get the appropriate airplane flight manual and study the following systems: engine, fuel system, flight controls, pitot-static, and the electrical system. At the end of a reading, you ought to have a clear idea of such matters as whether the air-driven gyros operate by vacuum or pressure, which panel instruments are air operated and which are electrical, the source of alternate air to the engine induction system, alternate static pressure sources (if none is mentioned, you can obtain alternate static pressure on any airplane with a vertical speed indicator by breaking the glass on the face of that instrument), and how to recycle a tripped alternator. The list is not limited to these, but some airplane flight manuals *are* limited, so it's difficult to make blanket statements about them.

With the manual close by, spend some time sitting in the left seat of the airplane, even if you've flown it often on previous occasions. Believe me, you are about to become ever so much more intimate with its workings, so time spent fitting the cockpit to you is worth hours in the air. Many experienced pilots who have flown more types of airplanes than a logbook can hold start out to learn each type by sitting in it and not leaving until they can accomplish every cockpit function without looking to see where the appropriate handle is located. It's a good method for starting out in any airplane with which you are unfamiliar or even just a little rusty. Pay special attention to circuit-breaker panels, fuse interchangeability in fuse panels, and fuel selector positions and function. The more complex the airplane, the more time you should spend on this exercise.

Once you feel comfortable with the physical layout of the cockpit, arrange to fly the airplane (VFR, of course) to become equally familiar with its responses. Pay particular attention to how the airplane responds to changes in power setting, whether translated in manifold pressure or rpm. You can approach this problem in different ways, and the one you pick is entirely up to you. One way is to memorize the numbers that produce certain conditions so that you can set, say, 20 inches of mp and know that on this airplane, you'll descend 500 fpm at cruise airspeed. Another formulation for power performance equates an inch of mp to 100-fpm climb or descent. In other words, add an inch to level cruise power and you'll climb 100 fpm, subtract an inch from cruise and you'll descend the same amount. Those systems are fine if you deal with them as a concept and don't allow yourself to be hypnotized by a task. You may spend too much time looking at the manifold pressure gauge and too little flying the airplane. Sure, those pitch and power combinations result in known performance and you should be able to set them up in a way that doesn't distract you. Spend enough time in an airplane to know the *throttle position* — the way it feels in your hand — that produces the required performance, then use that as a foundation upon which to build adjustments. Just as you wouldn't fixate on any other instrument, don't fixate on the engine dials. So know the power settings, but get to them by feel and adjust with a glance.

Learn the airplane's out-of-trim pressures in various configurations so that you can rely on these pressures as an index of performance. Trim pressure changes with changes in airspeed, given a constant center of gravity (CG). A particular trim setting may result in very different *attitudes*, however, and that depends upon the power setting. Thus, you can descend on an ILS at a particular airspeed, reach decision height, add go-around power — and the nose comes up to a climb all by itself! You never touch the trim. Alter the loading CG of the airplane and the trim tab position will be different for all conditions. What you want to avoid is making too many trim adjustments, particularly during transitory maneuvers. For example, a turn may require some added back pressure, but only for a few moments, so you'd only be making trouble for yourself by trimming off the pressure in a case like that. Most lightplanes in which you are likely to begin learning instrument flying can be handled easily when they are well out of trim. You may observe a pilot flying a heavier twin who trims early and often, and you may be misled into thinking that's the proper technique for IFR. If new instrument pilots do one thing too much, it's trim adjustment. Pick a reference attitude and trim *away* the pressure, just as you were taught in VFR flying.

By the time you arrive for IFR lessons, you should be able to fly, almost by habit, the five configurations of an airplane that apply to instrument flying: climb, cruise, en route descent, approach, and approach descent. If you can get a trimmed-up airplane into each configuration smoothly, transition properly between each, in a minimal amount of time and almost by instinct, you can cut about three to four hours off your basic instrument training. Money in the bank. The best part is that you can do a lot of that practicing on your own and in VFR conditions by drawing yourself a power and configuration table:

	Airspeed	Power	FPM	Configuration
Climb	manual	manual	the more, the better	clean
Cruise	the more, the better	see table	0	clean
En Route Descent	same as cruise	cruise minus 5" or 500 rpm	−500	clean
Approach	about same as climb	close to en route descent	0	gear down, flaps 10-15°
Approach Descent	same as "approach"	"approach" minus 5" or 500 rpm	−500	same as "approach"

If you are able to establish those configurations and power combinations without a great deal of fuss and to move from one to the other smoothly, you'll have gone a long way toward practicing the manipulative parts of basic attitude instrument flying. You'll also have made life immeasurably easier on yourself if you can set up a very close approximation of a standard-rate turn without referring to the panel to establish the bank. Your goal is to use your panel scan as a *confirming and adjusting* glance to the manipulation you've just performed. If you have to tread among the needles just to initiate the maneuver, you're not in touch with the airplane yet. From the start, your instructor will be asking you to do things like, "Set me up a 500-fpm descent at 90 knots." If you have no idea how to achieve that, it will take some hours to teach you. You can learn that VFR and on your own time.

Simulators are another way to enhance your education by substituting a machine that's designed for teaching and usually less costly in the long run for learning procedures than an airplane is. Think about how much time you waste just getting yourself into the training context in an airplane. With a simulator, five seconds after you're seated (oh, that may be a little exaggerated), you can be plunging down a storm-tossed localizer with June Allyson on the ground looking anxiously skyward as you wrestle this baby down while dealing a hand of blackjack — in other words, you name it, the machine will give it to you, and right now.

There are "trainers" and there are "simulators." An instrument trainer is a device that produces instrument readings that duplicate flying conditions but without the accompanying motions and accelerations you'd experience in a real airplane. A true "simulator" actually banks to the left when you crank the controls over into a left turn. It's possible to argue the merits of either approach until the tower crew goes home, but allow one observation anyway: when airplanes turn, climb, or descend, you'll feel accelerative forces that are very different from those obtained by simply placing a pilot in a container and tipping the container in the turning direction. For one thing, centrifugal force is entirely absent. The merits of motion simulators are therefore open to considerable discussion. Instructors who work with both devices say that the most important design functions are control panel and cockpit layout realism along with true-to-life ATC procedures.

You will also find that each simulator or trainer "flies" with its own individual characteristics, and you'll have to spend just as much time getting accustomed to its responses to control and power changes as you would with an airplane. Whether an airplane or a simulator/trainer, every machine has its own individual personality, that peculiar combination of values for lift, drag, thrust, and weight that make it unique, even compared to the ones on either side of it on the production line. Among different types, the single characteristic that requires the most adjustment by you, the pilot, is the designed-in drag. Very clean airplanes can attain high airspeeds in descents very quickly, and you'll have to learn to anticipate that when you fly them. Draggy airplanes are easier to fly in that respect, because they aren't as anxious to get ahead of you, and it's easy to adjust descent gradients. In the future, we'll see demand for ever greater efficiency, though, and that equates directly to lower drag.

At some point, you will have to decide how you will go about your formal instruction. Some schools offer a comprehensive program that combines a ground school with simulator/trainer exercises and all course materials thrown in. To save some money, you may wish to study for the written test at home using a test guide or a commercial home-study course, then follow up with flying training at a local airport flying school. You may even elect to use your own airplane and train with an independent instructor certified to teach instrument flying (and you may want to check to be sure of that).

Whatever avenue you choose, you should schedule your IFR training so that you can involve yourself in it *daily*. Weekend training may suffice for obtaining a private license, but it is not an efficient way to get an instrument rating. There is simply too much slippage between lessons, and you'll end up wasting time and money, possibly becoming frustrated at your slow progress and perhaps even abandoning your goal. More and more pilots are seeing the instrument rating as important enough to merit special preparation and even the use of a two-week vacation, say, to devote entirely to full-time study for the rating. There is no shortage of schools to cater to that type of student, and they are advertised widely in aviation magazines and newspapers. If you can't swing that sort of arrangement, discuss with your local school a schedule that will afford you the maximum continuity once you begin. Much of that will be up to you, of course, but if you make it clear at the outset that you wish to pursue a concentrated program, you'll avoid disappointment later on.

Selection of an instructor is even more important than the school. You can walk into a building with fancy furniture in the lobby and the most modern instructional equipment and still encounter a mean bunch of instructors with low morale because of some fundamental gripe with the school. If you get stuck in this kind of situation and realize it only after you've started, quit before too much damage is done and go find a place that suits you. Get used to the idea that you're paying for this. Too many schools try to intimidate a student by talking as if they're practically an arm of the government. You are purchasing a service, and you are entitled to professional, competent instruction. An instructor who strikes you as even slightly less than interested in seeing you succeed at achieving your goal is a risk. You want someone who feels positive about his or her job, about your decision and motivation to seek a rating, and about the ATC system. Be suspicious of any negative expression in any of these areas. An instructor who asks you why you want to get an instrument rating as is as much as telling you that there is some question in his or her mind as to whether you *should* get one. An instrument instructor ought to feel that every pilot who desires to obtain the rating ought to get the best possible chance to succeed. This is not to say that every pilot will succeed or that instructors should pass everyone even if they perform poorly. All it says is that an instructor should have his or her own personality and emotions well enough in hand to teach instrument flying without prejudice or value judgments about a student. Nobody can tell you how to spot the "wrong" instructor; you'll just know. The good ones are usually pretty well known by word of mouth, so ask other students who've already begun training.

Last, in IFR training, as with other things in life, you generally get what you pay for. This is not to say that a high-overhead slick operation is the best for you in terms of value delivered for your training dollar, but the low bidder is not always the best choice either. Be wary of schools that are engaged in a heavy amount of contract training because experience among students has shown that these organizations tend to be less attentive to the individual student; they don't mean to be, but it just happens.

Some flying clubs have instructors associated with them, and in general, the cost is lower than the equivalent fixed base operator (FBO) or school in the same area. It is impossible to come up with any hard and fast rules that will ease your decisions on the question of quality instruction, and that pertains to the clubs as well as to any other teaching facility. The particular affiliation of the club may affect your decision, but then again, it may be their equipment. You may just like the instructor.

You'll find an array of services that provide weekend cram courses for the written exam. They do a good job of preparing you for the exam, but I think you short-change yourself if you don't prepare on your own, either through a class or through reading, before you take one of these quickie courses. Remember that the written is nothing compared to actually *knowing* the rules and the system, and getting a good mark, even with little actual study and effort, may not leave you feeling very confident to go out and fly later on. Study and learn, then get prepped for the exam.

And now a word or two about the Slump. Nobody likes to talk about this, but it is part of learning anything, especially instrument flying. The *Flight Instructor's Handbook* calls this "reversal of progress," which is the polite phrase for the Slump. In the last chapter we talked about a curve called the asymptote. So that this chapter can have a curve in it too, we now present the *learning* curve. See that flat part there right after the initial great leap upward? That's the Slump, and unless you have the mind of a computer, it will happen to you.

Dedicated instructors say that a teacher who's sharp, experienced and tuned in to a student can prevent it from ever happening in an average learner, and that if it does happen, it gives the instructor a golden opportunity to talk about something that occurs in real-life IFR too. You have to learn *now* to handle a bad day, in other words.

Basically, here's what it's like: you'll be going along at a normal pace, maybe even beginning to wonder what all the fuss is about. Hell, this instrument business is really pretty easy, and you kind of wish that the person who sits in the right seat during your lessons would let you move along a little faster. All of a sudden, one morning you'll begin a lesson and nothing will go right. You start to sweat and your mind races. You try to concentrate, but that seems to make things worse too.

If you're human, you'll have a day like that, and it's nothing to fear or feel ashamed about. What's happened is that you're not putting together all the knowledge and skills you've learned in a coherent manner. *You* may think you are, but you're really not. You've lost the ability to integrate all the individual tasks that make up an IFR flight, and instead you're focusing on one task, then

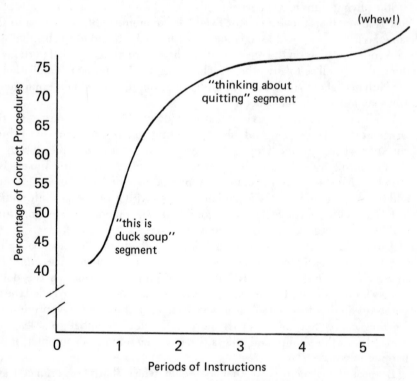

Typical learning curve (or, why me, Lord?) (Cribbed from the FAA Flight Instructor's Handbook)

another, in sequence. It's an unconscious lapse, which is why it's so bedeviling, and one of the best things you can do is to call off the lesson and land. If this sort of thing keeps happening, obviously something else is seriously wrong, but as an isolated incident, just file it away in your memory and start fresh tomorrow.

And the word "tomorrow" can be taken quite literally. One of the advantages of training for your instrument rating on a daily schedule instead of on weekends or at scattered dates is that you can recover from a bad day immediately instead of moping about for a whole week worrying about it. You've heard the saying, "You have to get back on the horse." It's true. If your instructor is any good, this will come as no great surprise to him or her. Most teachers have been through this themselves and have watched students go through perfectly wretched days in the midst of steady upward progress.

The Slump is a failure, make no mistake about it. It's no joke, and once you get back on the ground and tied down, you should spend some time meditating over the causes, preferably by yourself and in the left seat of the offending airplane. Bad days are a private thing between you and your instructor, and nobody else will know. Just think about how lucky you are that you weren't all by yourself when everything fell apart, because after you get your rating, your instructor can do nothing but read about it in the papers.

Now look at the learning curve again. After that flat spot, it goes back upward. Glad you got past it, aren't you?

IFR Notebook

Requirements for a pilot to receive an instrument rating:

1. A valid pilot certificate
     private pilot — 17 years old
     commercial pilot — 18 years old
2. A valid medical certificate appropriate to the pilot certificate and the privileges exercised, i.e., second class for commercial use of a commercial certificate, third class for private use of *any* certificate
3. Read, speak, and understand English
4. 200 hours total flying time, including —
     100 hours as pilot in command
     50 hours cross-country in category (example: airplanes)
     40 hours of flying time on instruments
     15 hours of instrument instruction, at least five hours of which must be in
          an airplane
5. A passing grade of 70 or more on the instrument written exam, earned sometime in the last 24 months before the flight test
6. A logbook record of instruction and competence
7. A statement from the instructor attesting to instruction in required areas, no more than 60 days prior to the flight test

About these rules:

Up to 20 hours of the 40 required hours of instrument time can be taken on an *approved* simulator; check with the school or FBO to ensure that such time is creditable. Your local General Aviation District Office can advise you on what types of simulators will be acceptable.

Instrument ratings are issued for *categories* of aircraft: airplane, rotorcraft, glider, lighter-than-air; and therefore, your pilot certificate must be issued for the category in which you seek the instrument rating.

Whereas the definition of cross-country is stipulated very clearly for student pilots (beyond a 25 nautical mile radius from the point of takeoff), the definition is not quite so clear in other portions of the FARs. According to some interpretations, therefore, you may count as cross-country time any flight that lands at an airport other than the destination airport, regardless of distance flown, for the purposes of instrument rating qualification.

As only 15 hours of the 40 hours of required instrument time need be instructional time, and only 5 hours of that need be flown in an airplane, it is theoretically possible to spend 25 hours of the 40 required instrument hours in solo practice with a rated pilot aboard as a safety measure. Furthermore, the time you spend under the hood or in a simulator during your training for a private license (assuming you insisted upon integrated training or the school offered it) also counts toward the 40-hour requirement. Although the regulations may make it appear that you could short-cut even further by flying a simulator *solo* for 20 hours, such a reading of the regulation is completely unrealistic. Do not plan your finances for an instrument rating based on such a reading. You must have an instructor's statement of competency in order to receive a rating, and it is taking something like 50 hours plus these days, on a national average, to become competent for the airplane instrument rating.

Instrument time should be logged according to the conditions in which it was flown. Some older logbooks are divided only into "airplane" and "simulated," which is not particularly useful. Time flown in an airplane with a visibility-limiting device such as a hood is supposed to be logged as "simulated" as opposed to "actual" time. "Actual" instrument time means time flown in IFR weather conditions. Most newer logbooks have a separate portion for logging ground trainer time as well ("Link" or "Trnr"). Remember that actual instrument time on almost any flight will be slightly less than the total duration of the flight, since it must terminate with a brief visual period before landing and roll out.

Military pilots seeking a civil instrument rating may be able to get one with just a written test (no flight check) if they meet the requirements of FAR 61.67. Pilots holding non-U.S. pilot certificates with instrument privileges will be issued a U.S. instrument rating (which is needed to fly IFR in the United States) if they first pass a special test covering Subpart B of FAR 91.

An airline transport pilot rating for airplanes includes instrument privileges; no supplementary rating is necessary. An ATP for helicopters may be limited to VFR only, however.

CHAPTER III

The Soul of Machinery

Preflighting is a subject nobody really likes to talk about. In a very basic sense, preflighting an airplane is a ritual, almost religious in nature, that calms the very reasonable fears about the mechanical nature of any flying machine. Pilots are limited by the rules in what they can do to an airplane. If I had my way, before every flight, particularly IFR flights over nasty terrain, water, or cities, I'd pull a compression check, do an ignition analysis, carb adjustment, prop governor overhaul, and a number of other procedures that would practically guarantee the successful outcome of the flight. But I can't do that. I'm not a licensed mechanic.

So we pilots are reduced to our ritualistic dance around the airplane, plucking at hinge pins and eyeballing orifices, all in the hope that this procedure will ensure that nothing will go wrong. Anyone with any sense knows that our relatively superficial inspection will probably never approach detecting a hidden engine malfunction that is four hours from happening or even a fraying alternator belt, the way some engines are cowled. Loss of a communication radio constitutes an emergency (in ATC's book, anyway) on an instrument flight, but our preflight routines allow us no tools or methods to check the health of those boxes. So let's face it: to say that we've gone as far as we can go to protect our own safety and that of our passengers by conducting the standard walk-around preflight of an airplane is to kid ourselves and everyone else.

First of all, not all airplanes are the same. There are new airplanes, airplanes that are broken in, airplanes that have been repaired recently, airplanes that are just plain worn out, and airplanes we don't know a damn thing about. Would you preflight each one of these in the same way? Of course not, and before flying IFR, you must discard some of your notions about airplane systems. An airplane has a finite life, as does each of its component systems. Your goal is to make sure that most of your flying occurs before one system or another dies, and in order to succeed at that, you have to look at the airplane in terms of rates of change.

Chances are, the airplane in which you learn to fly IFR will be your basic standard aluminum lightplane. When it was brand new, there was literally no wear on any of its parts, but as it ages, different parts of it wear at different rates.

The airframe itself wears out, though at the slowest rate of any component. (This is not always the case. Some very large transport airplanes in military service have wing spars with a rated life that sounds more like a lightplane engine TBO.) The metal frame is designed to bear certain maximum flight loads, and it flexes when these loads are imposed. Flexing metal produces "fatigue," usually in a fairly direct proportion, with a rapidly increasing rate of deterioration just prior to failure. Mind you, most lightplanes fade into obsolescence long before their lives end due to airframe fatigue, so for all practical purposes, there's little point to conducting spar inspections prior to flying.

Certain types of airplanes have remained in service long after anyone might have guessed their useful lives would be over, however; the Twin Beech and DC-3 come to mind immediately. Both airplanes are well designed, obviously well favored by pilots and operators, but they present a very different case to a preflighting pilot than would, say, a year-old Skylane. Both the older airplanes carry a lot of cargo, which means they fly heavy a lot, probably on service that's at least regular if not scheduled. This tells you the airframe has had plenty of time to flex and develop the symptoms of age. Since we are talking about a very slow rate of change in the case of airframe wear, a long interval between your inspections of the airframe is perfectly reasonable, but if you are flying an older airplane, you should make it your business to have a look at its innards when such an inspection is convenient. At the very least, every airplane must receive an annual inspection, and with an older airplane, it is worth your while to arrange to be present for a visual inspection with the help of the airframe mechanic. He will show you where to look for any signs of change, should you decide that more frequent inspections are necessary.

Floatplanes and seaplane/amphibians that fly IFR also age at a rate much faster than landplanes because of the extra loads they must bear in water landings and takeoffs, and visual inspections of the airframe ought to be part of your preflight routine with them.

Different parts of the airframe can be expected to wear at different rates. Certain parts flex while others move within hinges or turn on bushings or bearings. Obviously, the wings bear a load that flexes upward on them. The horizontal tail components bear a load *downward* to counterbalance the airplane and produce stability (in a conventional airplane). Load-bearing components withstand different types of wear from different sources. Gusts produce very high transitory peak loads: the turbulence inside a cloud, for example.

Vibration affects all parts of an airplane, even in the most hidden recesses.

Vibration has little to do with loading. Load stresses are low frequency, high amplitude movements that deliver very obvious results — they may bend something. Although load stress can act in a way that's very similar to vibration, vibration may act on *non*-load-bearing members, such as engine baffles and cowling pieces. Sound waves are doing the flexing here, at a higher vibration and lower amplitude but with destructive effects that can appear in a much shorter period of service life than do the effects of load.

Hinges and other moving surfaces can be expected to wear, obviously, because they rub against each other when they move. Most lightplane hinges are connected to ailerons, elevators, trim tabs, and rudders. It turns out that worn hinges are pretty important because they have a contributory role in something called "flutter," which is as nightmarish a situation as you can imagine. Flutter occurs at very high airspeeds when an aerodynamic surface vibrates until it is destroyed, or at ordinary airspeeds if it's out of balance. Obviously, a loose portion of a trim tab or control surface hinge just loves to flap around and vibrate. Hinge wear occurs over a long period and can be easily caught by periodic service inspections. There is no good reason to preflight an airplane without checking for looseness at hinge points, however, unless the control service is out of reach, in which case you should make it a point to use a ladder and check it every four or five flights, say, or anytime it's on the wash rack.

You can't see control cables and interconnects, but you should be able to feel any sloppiness or out-of-trim condition when you fly the airplane. An airplane that's out of rig — one in which the adjustments for control cables or interconnects are loosened or moved out of alignment — is so unpleasant to fly that few people would let the condition go without service for very long. Fuel tanks are also hidden — either separate rubber bladders built into the wing and fuselage or a section of the wing, inboard, that's been sealed with a rubberlike compound to contain liquid fuel. Since wings flex and the compound changes with age, sooner or later such fuel tanks will develop leaks. On low-wing airplanes, you may never detect such a leak simply by watching for puddles on the ramp. You must peer under the wing and watch for stains from the dye in the evaporated fuel, either bluish or red in color. Any fuel leak can become enlarged at any time, once flow starts, so the first sign of a leak calls for caution in any further flights to avoid long legs between landings. Head for a repair station.

Vibration cracks are not always visible, and since it's not practical to disassemble the airplane prior to each flight, you should do the next best thing and more closely examine areas where cracks are likely to develop in nonstructural components. Engines and propellers are a wonderful source of noise at just the right frequency to do this sort of damage, so start looking there. Sheet metal will tend to continue cracking once it has started. Fiber glass is difficult to crack at all, and the fibers in its structural matrix tend to stop cracking or to slow it down. Plexiglas canopies are like sheet metal in propagating cracks; they have to be drilled out to stop any crack from spreading. The propeller itself flexes along the length of its blade, and any imperfection in the leading edge is a possible source for a developing crack.

Older plastic components are also subject to a long-term destructive effect called crazing that produces filamentary lines in the surface of the material. In canopies, it can be blinding in strong sunlight, and it marks the beginning of the end for any component. Vibration is alive and well anywhere in the wake of the propeller as well, so observe parts that live in the buffet. If you want to experience what it feels like, carefully place a finger into the airstream sometime when the airspeed is down around 70 knots or so. It's surprising.

Antennas have it particularly tough, because they are like great tuning forks living in a world of vibrating air. I have never had one fall off, but I've heard of it often enough, and any preflight check should include a quick grasp of the key antennas to ensure that they're not just hanging on by the cable connector. Antennas are supposed to be mounted with plates called doublers beneath the airplane's skin at their bases to reinforce the anchor point.

On an older airplane, you may reasonably expect to find a number of areas on the airframe where doublers and other sheet metal repairs have been made. Chances are that those are the areas in which cracks will develop in the future as well, so concentrate on them.

Some airplanes are cowled in such a way that you can open only the oil dipstick door and peer through the cooling air inlets; the rest is latched in such a way that uncowling the airplane before each flight is neither practical nor expected. In fact, uncowling such an airplane repeatedly may wear the latching or fastening devices. If you are willing to allow all of that to prevent you from gaining access to the engine compartment, then you are insufficiently assertive. Even if you postpone under-the-cowl inspections to every other flight on such an airplane, you are doing so with the knowledge that the engine and its related equipment has a much shorter service life than the airframe.

We measure the airframe life of a lightplane in years and the engine in hours, usually 2,000. This is not to say that a 200-hour-per-year airplane will not give you 10 years of engine service; it could. What it does say is that a rental airplane that puts on 500 hours a year deserves your attention.

It should encourage you to know that engine failures are usually not sudden. Engines fade rather than drop through some figurative trapdoor. The reason the term "engine failure" is so commonly associated with what is really *sudden* engine failure is because the latter is so much more dramatic. The gradual fading of an engine is to be preferred to the sudden kind, but both modes are failures; one simply occurs at a faster rate.

The fundamental measure of the health of any engine is its cylinder compression. Without compression, the engine can't do any work, even if it manages to turn over. Without compression, there is no torque or horsepower, generally in that order, and the performance of your intended airplane will disappoint you mightily. Compression is related to the sealing of the piston and cylinder; it can involve the piston rings, the valves and their seats, and the seal at the cylinder-to-cylinder head junction. Any number of factors can affect all of these points. Since you can't preflight an engine for compression, you must rely on the periodic inspections by a mechanic. Other symptoms betray a low cylinder, however, and if you're alert to these, you can size up a compression drop before it places you in jeopardy. Low compression shows up most readily at low rpm. The engine will be difficult to start if all the cylinders are evenly worn, as might be the case with an older engine that needs rings. An uneven idle that won't respond to treatment (plugs, wires, ignition, carburetion/induction checks) may signal one low jug. While low compression is not necessarily a sure sign of sudden engine failure, uneven running among cylinders at high power settings can lead to a chain of internal events that would nauseate any mechanically sensitive person. If you could *see* what was happening, you'd ground yourself.

If there is anything at all surprising about engines, it's how much mistreatment they can handle and still keep running. The classic ham-handed engine torture method is to heat the whole mechanical circus up in a climb, then yank the throttle closed to descend in a screaming arrival at high airspeed. Alternately heating and then cooling the engine abruptly causes it to expand and contract in such a way that parts of it may warp. Gradual heating, such as occurs on a normal climb on even the hottest day, does no harm at all; but sudden chilling of the assemblage of parts will probably guarantee that some will shrink at a faster rate than others. The results are warped heads and cases, cracks, and loss of compression.

You'll hear pilots talk about valves all the time, but if they were speaking more accurately, or if they knew better, they'd really talk about valves and seats. It takes both to seal the cylinder during the power stroke, and a valve seat can be made to leak with just the slightest bit of dirt or mechanical force that distorts it.

As an engine ages, gaskets and O rings — for that matter, any sealing surface of rubber or plastic — will probably dry out and lose their sealing capacity. Although this may mean a messy engine and loss of oil, it usually doesn't directly affect safe operation unless you've got yourself a real gusher on your hands. The damage done by leaking oil is long-term. Oil causes rubber insulation and hose to rot, shortening its service life. Fly an airplane with an oil-soaked engine, and sooner or later you can look for other complications. The gaskets that seal exhaust and intake manifolds seem to have a limited service life, and their ultimate replacement is both expected and routine. You replace these as it becomes necessary. Exhaust seal failure is betrayed by streaks of grayish stain around the small openings near the course of the exhaust circuit under the cowl. Cooling air may force the exhaust gases out anywhere, but they are sure to condense on the exterior paintwork and tell you well in advance of any upcoming surprises. Leaky exhausts can blow carbon monoxide into the cabin, although the threat of that is overdone. Leaky intakes can result in lean running cylinders, never rich, which can mean burned valves and detonation.

Changes in engines occur in a predictable manner and over a fairly long period of time. Accessories mounted on the engines are something else again. Since aircraft carburetors are so basic, very little goes wrong with them at normal power settings. Carburetors can be installed with poorly designed intake systems, however, such that the mixture control at idle or very low taxi-speed rpm is very poor. You may actually be running an engine in a seemingly harmless low-power condition and fouling a spark plug or two so that on takeoff, you'll be greeted with a wonderful, rousing roughness that will cause you to awaken instantly and reconsider the entire enterprise of aviation. Manage mixtures at all times — ask to see your plugs when they're pulled or remove them yourself after learning how — and you'll avoid such momentary scares.

Magnetos are a part of aircraft powerplant systems because nobody has taken the trouble to certificate anything else yet. They are a very old-fashioned way of supplying spark to an engine with no requirement for anything other than crank motion to turn their windings, which is their advantage over an automotive-style battery-distributor ignition; these require an ancillary source of current. Magnetos also fail gradually under ordinary wear, and I've never been able to find any common denominator to the failures. You detect failure during magneto checks on the ground and during flight. Magneto checks in flight over a good spot for an unplanned landing is an excellent habit to get into, if only because it means you realize that systems operate *all* the time, not just during the moment when you were taught to check them. Expect a magneto to fade on you at some point if you fly long enough and often enough, and that way it will come as less of a surprise to you when it does. You will probably not even notice it in ordinary flying conditions.

Alternators come from the same factories that make the same piece of equipment on your automobile, but on airplanes they are exposed to greater vibration and temperature extremes. The only parts that wear on an alternator are the belt connecting it to the propeller shaft or crank and the bearings that take the side thrust of the belt — and that's *all* they can take. Alternators actually produce alternating current, and this is turned into direct current for the airplane systems by a circuit made up of solid-state diodes. Failure of a diode is usually the cause of any sudden alternator loss. You cannot preflight for that, either, and there is no way to get the alternator back on line if it happens. Some alternators last for twenty months and others for twenty minutes. They seem to have a lousy sense of rhythm that way, and maybe that's why nobody likes them very much.

Vacuum pumps are as mysterious as alternators in the way they tend to fail.

You can try to keep the filter end of their air system as clean as possible, but beyond that, there is nothing you can do during a preflight that will detect a vacuum pump failure that has been "scheduled" for later that day. Sorry, but these are the conditions under which you sign on.

So the engine itself is a sort of orchestra whose individual musicians may walk in and out at will during a number. Still, the times when the music stops altogether are very rare, and you shouldn't be distracted by the over-publicized catastrophic engine failure. Accustom yourself to the idea of engine failure as a very slow, long-term mode and you'll be in tune with the way things really are.

As I hope you're beginning to see, preflighting an airplane is not something that's limited to the five minutes prior to a flight. It really means awareness of whole systems, how they work and how they are likely to wear — and *at what rate*. Think as a statistician does; when you can predict engine condition, you'll improve your chances of making it through a long career of IFR flying with nary an incident. As a friend of mine puts it, "Engine failures should be, at most, an inconvenience, never an emergency."

In IFR flying, a reliable source of electricity is of paramount importance. Lose your battery and alternator on a VFR flight, and you can simply shut everything down and continue to your destination, landing with light signals if necessary. You might expect the electrical system to be nearly fail-safe, since it has so few moving parts, but that assumption turns out to be overly optimistic. Whereas the other systems in an airplane present you with predictable rates of wear and a kind of descending hierarchy of expected service life, from the durable airframe to the somewhat less durable engine to the slightly quirky alternator and vacuum pump, electrical systems don't fail because of wear so much as for reasons that sometimes appear magical.

Aside from the battery, which has a predictable service life, electrical systems in airplanes may fail very early (something known as "infant mortality") or chronically, owing to some defect that, because of the nature of electricity, simply eludes detection. Airframes and engines are hardly ever described as failing "intermittently," a talent that only electronic components seem to have. An electronic flaw can be the most difficult fault to track down in an airplane, and even after you assume you have located the source and repaired it, it may reappear. In the context of preflighting according to our method of organizing the airplane into systems with an expected service life, the electrical system and electronic components such as avionics simply don't fit. They defy the rational approach.

Batteries require far more attention than most operators give them. Part of the reason for that is their relative inaccessibility in corrosion-resistant cases tucked away in some remote corner of the airplane. Lead-acid batteries wear through a process called sulfation, which allows a precipitate to collect in a reservoir in the electrolyte below the plates of each cell. Once this compound gathers in sufficient quantity, it can bridge the plates themselves and form a high-resistance short circuit. When that happens, the remaining electrolyte is heated and boiled away. It is the beginning of the end for the battery. You can delay the onset of this wear by using the cleanest and softest water possible, preferably distilled water, but aircraft batteries are very compact for the work they have to do, and their service life is short. You should begin to check your battery more frequently and with greater attention once it has been in service for about a year, for average aircraft usage.

But the battery is not intended to be an aircraft's primary electromotive force — not anymore. Its main job is to supply starting power to the engine, with the alternator taking over after that. Therefore, the battery serves the IFR pilot as the backup power source in the event of total alternator failure, and therefore deserves to be regarded as an emergency system only. If you have the attitude, "Oh well, if the alternator fails, there's always the battery," you're taking a chance. Remember that batteries deliver much less than their rated power when they are cold, and an IFR airplane with an electrical emergency is not usually at room temperature.

An airplane with a persistent electrical problem is a risk in actual instrument conditions, and there is no way to avoid that conclusion. Seek service for the problem immediately, and after the flaw has been repaired, fly the airplane VFR for some reasonable interval to find out whether the fault has been fixed. Any intermittent failure is particular cause for extra caution, since you must assume that it will occur at the worst possible moment in the next flight.

If every pilot had some servile being who existed only to announce events like, "Your flap link on the right side will part company with the actuator in 2.3 hours," they would immediately have the defect taken care of. If someone told you you will have an electrical failure within 43 minutes of departure, you would not depart at all. But there is no such voice to advise us. We have only ourselves. And if we persist in regarding an airplane's mechanical and electrical essence in the detached way that typifies the traditional preflight check, we give ourselves no advantage at all. You can't become a mechanic, but you can think like one. Nobody expects you to make the actual repairs. What you ought to expect of yourself is a sensitivity to the airplane and its systems. Instrument flying demands it.

New airplanes pose some interesting risks of their own, and you should challenge any notions you may have that because an airplane is new, it will be trouble-free. The fact is that most experienced pilots are very cautious with a new airplane because errors that may have been made and not discovered at the factory will probably show up in some form of problem very early in the airplane's service life. The "infant mortality" phenomenon is not peculiar to airplanes; it is true of every machine ever made, including something as complicated as a hatchet. Remember that every new airplane presents a unique problem to the designer. It is as if he or she is inventing the airplane all over again from the beginning. Don't be so shocked, therefore, when that designer, who is only a human like you and me, in reinventing the airplane, learns something about part of that design after the airplane has been in service for a couple hundred hours. Experienced pilots with new types may get exasperated at the down time that may follow, but they know that any new machine has a period of debugging, and they've come to expect it.

Preflighting involves really getting to know an airplane, and rental airplanes present a genuine problem. Other people fly the thing when you don't, and they may do things to the poor machine that may victimize you at the least desirable moment. Rental airplanes should never be flown into difficult IFR conditions until you have flown that airplane for a considerable period of time in visual conditions that would at least allow you to recover from any system failure. If the airplane's service record is available for you to examine, ask to see it. Look for repeated failures of any system, and try to find a pattern in the failures. If it appears that someone is merely replacing the part and failing to deal with the fundamental flaw, don't use the airplane either VFR or IFR.

There is no real substitute for owning your own airplane when you fly IFR regularly. With ownership comes a certain peace of mind about the airplane itself. Even a limited partnership is better than rental. If you must rent, stick with one airplane that you know well. Fly it long enough to learn its mannerisms before you take it out IFR.

Once you have an airplane with which you feel comfortable, that you know well, you can organize a preflight routine that takes into consideration its peculiarities, its age, and its mechanical history. You can't write checklists for that sort of thing, but you must find some way — invent your own, and you'll remember it better than if this book gives you a way — to keep track of every system and peer into it at an interval that matches some reasonable expectation of its rate of wear. Treat any airplane that has just come from being serviced as if it were entirely new. If you don't see the wisdom of that, ask pilots who've made forced landings right after something as simple as an oil change. *The least predictable element in this whole equation is humans.* If you are perfect, then disregard all of the above.

What about the human in the pilot seat? We've just analyzed the airplane as a collection of systems with different rates of change. No airplane has the horse's talent, to fly you home to the barn unassisted. The pilot is part of the mechanical circuit and ought to be examined as closely as the airplane. Trouble is, most of us demonstrate our worst judgment when it comes time to evaluate ourselves. I told you to write your own preflighting system for the airplane you plan to use, but if you're like me, you can't be trusted to apply the same stern rule to your self-inspection. Here is a list of pilot "systems" that need careful going over before each flight:

Visual system:	Eyeglasses or contacts? A spare set?
	Sunglasses (glare can make adjustment difficult on a descent into dark)?
Pulmonary system:	Oxygen aboard if high altitudes planned?
	Smoker?
	Cold or hay fever?
	Antihistamine medication?
	Hyperventilation tendency?
Cardiovascular system:	Regular exercise?
	Okay under stress?
Semicircular canals:	Tendency toward vertigo?
	Cold or infection?
	Eustachian tube able to clear?
	Earplugs or headset?
Digestive system:	Light meal?
	Sugar "rebound"?
	Alcohol? NO GO within eight hours
	Pills or tablets? NO GO for most

CAUTION: Be wary of medicines designed to be absorbed over a long period through the device of small capsules with time delay coatings. Recognize these by examining the capsule and finding tiny beadlike pellets inside. If the time period is not specified on the package, phone the pharmacist and ask.

Control system (or, your Head):	Attitude
	Stress
	Fatigue
	Ego
	Age

This checklist contains most of the points outlined in different form in the *Medical Handbook for Pilots*, Advisory Circular 67-2, which is a very elementary discussion of aviation medicine. When we were preflighting the aircraft, we were limited because we are not mechanics. The same could be said of our personal preflights, since we're not doctors, but we got around the former situation by being reasonable, and there's only one limitation to our accomplishing the same with our own preflight physical: our judgment. Aviation has a sort of Catch-22 written into it: the only pilots who take advice are those who don't need it. If you fly with a known medical deficiency, you are not about to take the word of this book or anyone else. Ponder this, however: the FAA's Aeromedical Institute has observed that almost overwhelmingly, people who spend a lifetime fighting the loss of a medical certificate and finally regain the right to resume flying do not exercise that right once it's regained. In the Soviet Air Force, pilots are not entrusted with the evaluation of their own fitness. In the United States, the Soviet system of preflight physical examination is out of the question. There is no force to make you ground yourself. You are free to fly drunk, disorderly, and dyspeptic if you choose to. Just remember to add lots of money to your checklist, because you are going to need plenty someday. It might even buy you a friend.

An airplane with just a pilot in it is practically naked in an IFR environment, because instrument flying is nothing more than a very carefully constructed system. In order to fly in that system, at least from one point to another, you need certain equipment.

Charts are an obvious requirement, either the government (NOS) or the Jepp version satisfying it. They weigh so little that it's a shame to depart without the whole set. Wouldn't you feel just awful if some weather emergency forced you to fly somewhere you hadn't planned to and that part of your subscription was sitting at home? Bring 'em all. Also the sectionals. I've carried mine in backpacks, suitcases, briefcases, giveaway bags, cloth sacks, cardboard boxes, and everything but doggie bags.

When you first start flying instruments, you'll find yourself tempted to buy every piece of supposedly helpful apparatus you lay your eyes on. The psychology at work here is simple: you don't want to deny yourself *any* contrivance that may help your instrument flying. So get it out of your system. Buy the kneeboard that contains a miniature microwave oven. Buy the chart case that converts into a motorscooter. Believe me, the accessories industry can use the money.

Once you graduate past that stage, though, you'll find that you begin paring the junk in your chart case to the minimum. You will have found that the only valuable accessory on an IFR flight is a neat working area, and no accessory really provides you with that. You need something to write on—a clipboard with a plain pad on it is fine—and something to write with. To me, nothing can ever beat an automatic lead pencil of the type that feeds fresh lead to the point. Carry two of those and you'll never run out of ink, break your pencil point, or be at a loss when a new clearance arrives. Keep them in the same place all the time and they'll never get lost, either.

Carry a flashlight whether you think you'll be flying at night or not. Check it regularly for sufficient battery power, and replace the batteries every few months whether they need it or not. Batteries are cheap. There are also some nifty rechargeable flashlights on the market if the batteries aren't cheap enough. I used to use one until one day when I reached for it and remembered it was back home, inserted in a wall plug recharging itself. Don't leave the flashlight in the airplane, though.

Do keep a spare mike and headphone set in the airplane. Microphones and speakers seldom fail, but when they do, there is nothing quite like the feeling of staring at a perfectly good stack full of radios that you can't use. Losing your nav receiver is a picnic compared with the feeling when you can't talk to anybody.

An IFR-equipped airplane usually has a clock installed in the panel some-where, but airplane clocks are not always reliable. Many people spend hundreds of dollars on timing instruments of all kinds that split a second into thousandths and keep time better than the Naval Observatory. Do you ever wonder why they do, since timed approaches are so crude to begin with? If you think an expensive timing instrument — electronic or windup — will be more reliable than a cheap one, then buy the expensive one. But please don't do it because of accuracy. The accuracy simply doesn't matter.

You may also carry some equipment for survival in the event of an unscheduled landing on water or in hostile terrain. (Actually, the terrain is probably not hostile at all; we've just lost touch with how to deal with it.) They make some very compact "space" blankets that take up very little *space* in your airplane or flight kit, and they afford one of the best protections against the usual fatal element in any forced landing: exposure to temperature extremes.

You should also have a flight computer or calculator with you for that won-derful moment when the man says, "Radar's down, give me your estimate to Kansas City." I had an instructor once who actually said this: "Ninety percent of the time, you can just tell 'em fifteen minutes and half of that time, it *is* fifteen minutes." You want to make IFR as easy on yourself as you can, but that doesn't mean sloppy.

Munchies. Don't ask me why, but knowing you have something tucked away to nibble on is a big help. Unless you hate eating, a snack in the middle of a long en route leg can recharge your mental battery and break up the monot-ony. Anything you can do to enhance your own alertness is fair. Some people carry coffee, which has a mild stimulant effect but a much greater psychologi-cal effect along with a history of war stories about how it spews out of contain-ers at altitude. The third effect it has is diuretic — on some people, including me. It can be an absolutely insufferable distraction to instrument flying, so think about it before you drink too much.

American Cyanamid manufactures an emergency chemical light source un-der the trade name Cyalume; they're sticks that come to you packed in foil. To get light, you bend the unwrapped stick until a little vial inside breaks, mixing two chemicals that phosphoresce. The time of useful light varies with the tem-perature of the reaction, but these things seem to store almost forever and they give enough light to get you down quickly on a dark night. I like penlights and Sanyo's rechargeable Cadnica Lite, too, if only because you can hold them in your mouth and keep them trained on the panel while you use both hands to fly. I've never had to do that, but part of instrument flying is sitting around thinking about all the possible things that could go wrong. I have heard from very suspect sources that archaeologists excavating in Germany have come upon a fragment of manuscript that seems traceable to Georg Ohm. This missing part of Ohm's law says something which, loosely translated, reads, "If a lightbulb can burn out when you need it, it will." He was right.

We haven't even started to plan for this flight yet, and already there's a whole chapter full of things to worry about.

IFR Notebook

Required equipment checks for IFR flight in the United States:

1. In order to fly in *controlled* airspace under IFR, *each* static pressure system and *each* altimeter instrument must have been tested and inspected within the preceding 24 months to comply with Appendix E of FAR 43.

   Furthermore, no IFR flight in *controlled* airspace is allowed above the altitude for which the *altimeter* has been tested.

2. *No person may use* an ATC transponder that has not been tested and inspected within the preceding 24 calendar months to comply with Appendix F of FAR 43.

3. No pilot in command of an aircraft may allow the operation of any portable electronic equipment on any U.S.-registered civil aircraft under IFR, *with the following* EXCEPTIONS:

   portable voice recorders
   hearing aids
   heart pacemakers
   electric shavers
   any other device the *operator* of the aircraft has determined will not interfere with navigation or communication.

4. No *person* may operate a civil aircraft under IFR using the *VOR system of radio navigation* unless the VOR equipment of that aircraft is

   a) maintained and inspected under an "approved" procedure

   or b) has been operationally checked within the preceding *30 days* and found to be within permissible bearing error.

About these rules:

The requirement for testing of static pressure systems and altimeters every two years applies only to their use in controlled airspace. Notice that in airplanes with more than one system, each system must be tested. Altimeters can be tested *only* by properly certificated repair stations; an airframe mechanic can test and approve the static system, however. In each case, the shop should record the test in the aircraft log; it is a pilot responsibility to ensure that the test has been performed by checking such log records. There is no altitude limit for static-pressure systems, only for the altimeter instrument itself.

No *person* — that includes *anyone*: student pilots, passengers, juvenile delinquents — may operate a transponder that's not current with respect to the approved test. Only licensed repair stations may approve transponders, and the test should be recorded in a log record.

Appendixes E and F of Part 43 describe test procedures for altimeters and transponders, but since only licensed stations can perform these tests, the procedures apply only to them. Notice that manufacturers of new airplanes may obtain approval to test both altimeters and transponders, but be very cautious to note the dates when the tests were performed (or even *if* they were performed) in the logbooks that accompany a new airplane when it is delivered. Frequently, a significant period of time has elapsed between the date of manufacture and system approval and the final date of delivery to the end customer.

Very few pilots are aware that it is their responsibility to ensure, as pilot in command, that disallowed electronic devices are not operating aboard an IFR airplane. Remember that your insurance coverage may be affected if you should have an accident when some rule has been violated. You may feel a bit timid about asking passengers whether they are carrying any portable radio receivers, but you'd better be alert for it. Some passengers like to try handheld CB or amateur radio transceivers at altitude. The Canadian Ministry of Transport has also determined that handheld calculators may affect ADF receivers.

The old rule for VOR equipment used to specify a 10-day or 10-hour interval between tests, but the rule was broken more than it was followed. Now that the rule has been relaxed, there is almost no excuse for failing to adhere to it. If you plan a flight that will make *no* use of VOR equipment, you are legal to fly IFR without these tests current.

How to Meet the VOR Test

1. Use the test signal at airports that have one: ±4 degrees bearing error.
   This means that 4 degrees is the allowable error; a reading of 184 or 176 degrees means you are legal; same goes for 356 and 004 degrees.

The rule makes no mention of the "to-from" indicator, but you should check it:

0/360 From
180 To

Some people remember this by the mnemonic device "Cessna One Eighty-Two" ("180 To"). Another memory device is "north of the station," which is what the indicator says. Radio magnetic indicators register 180 degrees on a VOT.

2. Repair station test signals (avionics shops, etc.): ±4 degrees bearing error.

3. At airports without a test signal but with a published instrument approach procedure, look for a ground VOR checkpoint. Ask an FBO or the tower where the site is located on the airport. A complete and current list of test sites is published in the government's *Airport/Facility Directory*, in the back of the book. A ground VOR checkpoint is merely a designated location on an airport, usually with a small sign nearby, that has been surveyed and found to lie on a specific radial that you can receive from a nearby VOR transmitter. The sign will give you the bearings you should select on the OBS, and permissible error is ±4 degrees.

4. Use an airborne checkpoint, as listed in the *Airport/Facility Directory* list. The list stipulates the altitude at which you should fly to perform the check. Maximum bearing error: ± 6 degrees.

5. Find an airway that has a prominent landmark (the Regs say "ground point") on its centerline and visually confirm the corresponding radial with a maximum error of ± 6 degrees.

6. If two VOR receivers are installed and operating, you may check both by tuning the same station and observing the indicated bearings from that station with the needle centered. Maximum bearing error : ± 4 degrees.

If you value your hide, be very careful about using this form of test as a habit, since it is possible for both receivers to deliver serious errors with no independent means to check them. All the other test methods compare your equipment against some external standard. The dual test doesn't. Even if one receiver should fail the test using the other test methods, you can fly legally on the one that meets the maximum permissible error. If this test shows, say, a difference of 6 degrees between the two receivers, how will you know which one to believe? The answer is you won't. You might make a good guess based on known past performance, but you don't really know for sure.

You are required to record in some permanent log or other record the following data:

> date of test
> place and type of facility (VOT, etc.)
> bearing error
> signature of operator performing the test

There are several quick and painless ways to keep VOR tests current. One is to write the required test data directly onto the panel near the CDI with a soft lead pencil. If the panel material won't accept writing, install a patch of material that will. This way, every time you enter the airplane and so much as glance at the VOR, you'll see a reminder of when the test will be due. Another good method is to keep a small notebook as an "avionics log" and record the VOR tests there; the advantage of such a record is that you can spot long-term trends in VOR performance, and it will lead you to the more reliable of two indicators if they don't agree. Since VOT test signals are so easy to use, there is almost no excuse to miss checking the needles anytime you're near one of these.

It's not a bad idea to place similar placards for the required altimeter and transponder inspection dates where they are prominent and visible. I used to put stickers on the front of the airplane flight log as reminders for all the required inspection intervals and the dates when each piece of equipment would lapse. Since I recorded the flight data for each flight after shutdown, I was forced to glance at those dates regularly.

CHAPTER IV

Meet the Panel

Instruments come in panels. All those round, little faces can be so much more intimidating when they come at you in groups. Remember back when you knew next to nothing about an airplane, and all you could see was the awesome array of glass dials? Alienating, wasn't it? You felt like a nudist in a room filled with strangers. Gradually, you got to know some of the individuals on the panel during VFR practice, and the panel itself became less a whole entity and more an assemblage of functional servants, each with its own peculiar place and role.

Learning to fly on instruments requires only that you add to your intimacy with all the elements of this panel. There is, after all, very little difference between manipulating an airplane VFR or IFR. The laws of physics are unchanged, and you operate with the same three axes of motion and control — roll, pitch, and yaw — and the airplane will perform the same no matter what the visibility is outside. (The air may have different characteristics, but it's the same airplane.) When you flew from one point to another under visual flight rules, you used a chart and visible landmarks on the ground to ensure that you were headed in the proper direction. Instrument flying recognizes that you no longer are able to find landmarks on the ground. Instead, you use one of a number of electronic "landmarks" with known positions to tell you how well you're holding a course.

Those needles that dangle from the faces of navigation avionics are beguiling — you can't wait to use them. They seem nearly miraculous in some respects, able to tell you where you are when you can't see. Avionics can do marvelous things, but they distract too many pilots from the true functions of instrument flying.

The foundation of instrument flying is dead reckoning, using the panel instruments to manage the airplane over the short term. Overlying this primary function is the second instrument flying chore: avionics management and navigation, which tell you over the long term how to correct the primary function and at what rate to make the correction. Basic instrument panel flying comes first, navigation needles come second, and that is generally the order in which you learn to fly IFR too.

Make yourself a mental poster about three stories high and as wide as the Ritz that says this:

I Instruments — aviate
II Avionics — navigate

If you can get that fundamental idea straight, you'll have little trouble with instrument flying and you'll be part of a very small group of pilots who have come to grips with that significant truth.

A few observations about instruments: first, they replace your sense of up and down, which becomes unreliable in an airplane when you can't see. When you're standing on the ramp, your eyes and the pressure of your weight on your feet, modified by the data from your middle ears, tell you where the center of the earth is — straight down. Let's say you were to climb into the airplane and put that airplane smack in the middle of a cloud, then bend it over into a nice, brisk 60-degree bank. Where is "down" now? Your reason may tell you it's somewhere around eight o'clock low, but your senses are still saying "straight down."

Instruments also replace your sense of direction, for which you usually depend upon your eyes and your ability to think. Even on the ground, you can't find your way to some landmark if you can't see it. Since we change direction in an airplane by banking it to one side or another, any instrument that can tell us about bank angle and the change in yaw will keep us apprised of directional control from moment to moment. Over the long term, remember, directional control is called navigation, and that's distinct from the chore of maintaining direction right *now*.

We also need something to tell us about the rate of change of position. Another word for rate of change of position is *velocity*. Your middle ear (or, for that matter, any other human sensory organ) cannot sense simple velocity; it can only sense acceleration, which is the *change* in velocity. When you're riding in an airliner, you feel a sensation of motion at takeoff and during descent and landing, but that's it; aside from those transient periods of acceleration, you might as well be standing still, for all your body knows. As a matter of fact, the average middle ear is no good to you at rates of less than about 2 degrees per second per second in a turn, which helps to explain why pilots can enter into a gradual spiral descent in cloud and never know it.

It is also fun to ponder how instrument panels handle the three axes of motion, the "hinges" around which the airplane rotates. Pitch and roll axes are

The basic flight instruments: Top, left to right, airspeed indicator, artificial horizon indicator, altimeter. Bottom, left to right, turn coordinator, directional gyro, and vertical speed indicator.

depicted very clearly and graphically on the panel instruments; however, yaw is treated like a poor relative. At ordinary flight attitudes, a change in yaw usually means that the nose swings left or right — a heading change, in other words. Yaw is therefore usually interpreted through *direction* instruments whose primary function is to monitor the turn or rate of turn.

We're usually taught that the rudder is the single control surface that corresponds most directly with movement around the yaw axis, and the only instrument we have with a functional link to the rudder pedals is the inclinometer, or "ball," which is tucked away at the bottom of the turn-rate indicator. In normal maneuvers, we never turn the airplane using only yaw or only bank but a combination of both axes of movement. The predominant concentration is in bank, though, and most pilots perceive their directional instruments as displaying bank information and not yaw. Later in this chapter, when we talk about how each instrument works, the differences between what the instruments *sense* and what they actually display will be clarified.

In the functional sense, any VFR student pilot knows that pitch and bank controls provide a lot more freedom than do the rudder pedals — the yaw controls. Airplanes are fairly stable in pitch and roll, but given enough time without attention to the controls and they'll find their way into some increasingly divergent mode, like a spiral. If you regard yaw as an isolated axis of control, however, airplanes seem amazingly stable in that particular axis. Ever heard of an airplane switching ends? You have sufficient authority in pitch and bank to perform loops and barrel rolls, but it takes an exquisite rudder to generate even a tepid slip. Airplanes may roll over, but they hardly ever lose it around their vertical axis of motion unless you can get one to enter a spin. In a cruise configuration, however, with normal airspeed and the slightest attention to the rudder pedals, airplanes want to fly straight. When you fly under IFR, you will usually maintain the control over the yaw axis sufficiently to produce coordinated flight; control excursions in pitch and roll provide you with all the normal maneuvers. For IFR, just apply what you already know from VFR flight; control yaw with the rudder pedal by "stepping on the ball" — if the inclinometer slips out of its index to one side, add rudder pedal pressure on the corresponding side.

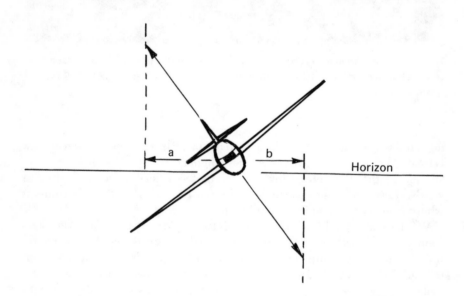

To center the inclinometer, or "ball," the force of lift must be tilted enough *into* the turn to form a horizontal component, *a*, that's strong enough to balance the centrifugal force, *b*. Instrument flying is no different from coordinated VFR flying: if the ball is off center, add pressure on the corresponding rudder pedal—"step on the ball."

There are various schemes for "organizing" the attitude instruments. Some books refer to primary and secondary instruments, others talk about pitch and roll groups, the rest break it down into control and power. The only thing they have in common is that none of these systems help you very much. What you need to know is how each instrument responds to changes in the airplane and how to apply what you know to good attitude flying.

All instruments in the conventional lightplane recognize the inherent stability of the airplane within certain limits and the fact that the IFR system is designed around very gentle maneuvering. A great deal of discussion centers around the errors inherent in instruments and their limits in roll and pitch. New students are exposed by ritual to wild tales about gyros that have "tumbled," the IFR equivalent of psychosis in a panel. The truth is that today's lightplane in good repair could be inadvertently rolled without tumbling the gyros. It could even be looped with no adverse effects other than the transitory meeting of the gyro mechanism with some limit stops; it won't "tumble," though. The technical standard order (TSO) for gyro instruments requires those improved characteristics and the major manufacturers of new, instrument-equipped airplanes use the TSO type component overwhelmingly.

All the instruments that tell you what's happening with respect to attitude references and velocity fall into groups according to their functional systems: *gyroscopic* and *pitot-static*. There is a third group that tells you about engine power and condition, and although they're close by and look just as important as the attitude flying group, they have nothing to do with manipulating the controls of the airplane. The power management instruments are no less important than they are to VFR flying, but since they operate the same way, there is nothing new for you to learn about them when you fly IFR. You'll learn to think about power management in new ways that show greater concern for rate of change when you fly IFR, and you'll find that your power changes are more precise and well thought out in advance, but aside from that, the manifold pressure, tachometer, and mixture monitors all do their same old jobs.

The gyroscopic instruments are based on a flywheel that's spun by some power source and maintained at a required rpm. The gyroscope depends upon two fundamental properties: its mass and its rotational speed. Affect either and you will influence the performance of the instrument. You can't change the weight of the wheel — that's constant — so that leaves only spin to worry about. You keep an eye on the spin by monitoring the power source that provides the energy to spin the flywheel.

The truly attractive talent of the gyroscope is its ability to cling to its original axis of orientation once it has been spun up to speed. You can orient the gyro any way you please, spin it, and it will stay put. Instrument designers choose the axis that will give them the best information when they design a gyro instrument. They know that a gyroscope can provide useful forces only in a direction perpendicular to its spin axis. Spin a gyro to its correct operating speed and try to move it through one of its two remaining axes of control and it will surprise you by exerting a force at 90 degrees to the force you impose on it. An instrument designer can utilize either the gyro's talent for rigidity in space or its ability to produce that sideward force (called *precession*) by using one or the other to present useful information that's meaningful to the pilot.

The gyros in lightplane instruments are spun either by a jet of air or by electric motors. The air-powered instrument may be driven by air under pressure or by air in a relative vacuum. Air-powered instruments are a little less expensive than electric ones, and they lack a flag to warn of loss of reliable reading (although that may soon change). Air-powered instruments have one inherent disadvantage in their erratic performance at high altitude, although the use of air under pressure improves slightly on air moved by a vacuum pump.

Air-powered instruments are generally reliable, except that they may become contaminated by their own power sources. Instrument repair stations say that although the newer filtered gyro instruments are much better than the old type that drew unfiltered air from behind the panel, some damage is still done by contamination. The source of contamination depends on the location of the air source. Intakes in the cabin pick up tobacco smoke; the ones placed in the engine compartment pull hydrocarbon fumes or exhaust leaks. Take your choice.

Air-powered instruments can't tell the difference between vacuum air and pressure air (pressure systems are a little easier to design and have the slight altitude advantage), and all you really have to know is where the air comes from and where it goes. After being filtered, it runs through lines to various instruments, and each connection where the two join must be carefully sealed. After the air has done its work spinning the gyros by striking turbine buckets in the gyro flywheel itself, it exits and is dumped into the atmosphere. In a pressure system, the pump that powers the air is upstream; on a vacuum system, it's at the very end of the line. All air-powered instruments need a certain pressure or vacuum value to operate reliably — call it a pressure difference. Some panels incorporate an instrument that may use a somewhat different value of pressure difference from that required by the others, in which case a flow restrictor may be installed to modify the pressure to match that particular instrument.

Air-powered instruments may lack warning flags, but the panel should provide a gauge with a green arc on its face to advise you of the power supply to the instruments. If it is at some distance from the main instrument scan on some airplanes, the air gauge may escape your notice, so you should include that in your thinking when you first start to fly.

Typical lightplanes drive part of the panel with air and part of it with electricity in order to spread the bets a little. It has become standard design to operate the attitude indicator and the directional gyro with air; the turn coordinator is usually electrically operated. Some airplanes may be set up differently, and you should learn all about yours in the most preliminary stages of getting to know its innards. The rationale behind splitting of power sources is that if either the alternator or the air pump checks out of the airplane, you will still have the remaining power source and its corresponding instruments. Naturally, if both the air and the electricity are lost at the same time, you'll have even less to work with — but don't think for a moment that you're doomed, because the airplane is still controllable.

Since gyroscopes merely spin on some bearings and otherwise move very little, internal mechanical failures are fairly rare. When a gyro goes sour, you first notice that something seems slightly out of rig. You can't seem to hold heading or the airspeed acts funny when you try to maintain attitude. Those are the first symptoms of a gyro winding down. Its precession forces or its willingness to remain steady in space are being reduced with time, and eventually it will be totally useless; you have a few minutes with it, no more.

You can spot cranky instruments during taxi check and also by noticing that they take a long time to behave properly after you power them up. An attitude indicator, for example, that is slow to erect from its usual sleeping position cocked to one side is probably giving you advance notice that it will perform less than perfectly in the air.

Pitot-static instruments provide altitude and velocity data by measuring air pressure. Each contains a device called an *aneroid*, a sealed metal box that distorts in a predictable way when you pump air into it or out of it, or in the case of airplanes, when the air pressure surrounding it changes with altitude. Magnify these distortions mechanically and attach a needle to the mechanical amplifier and you've built a simple altimeter. You can think of an altimeter as a barometer built around a different scale, if it's helpful.

If the aneroid leaks a little, it will distort only while the airplane is changing altitude, its amount of distortion being proportional to the rate of climb or descent. An aneroid with a precise leak is the basis for the vertical speed indicator, also called the rate of climb indicator. It's not a required IFR instrument but one that's become so widely accepted that you can expect to find it in any lightplane IFR panel. Both the altimeter and the VSI are connected to a static port, usually a small hole located somewhere on the airframe where its readings of direct pressure of the ambient air are minimally influenced by differences in airspeed. Finding a perfect location for a static pressure port can be tricky, and most airplane flight manuals list calibrated airspeeds that correspond to the normal operating scale of indicated airspeeds to compensate for the differences that arise. Usually, nobody bothers with that page. In general,

I like to look over the figures to see where they match most closely and where they diverge and by how much.

The airspeed indicator also contains an aneroid, but this one is used to compare the pressure of the air being rammed down a pitot tube with the static pressure of the ambient air at altitude, thereby compensating for the change of density in the air coming in the pitot as the pressure altitude rises. The only remaining compensation you must make in order to obtain true airspeed is a temperature correction. Some of the newer airspeed indicators have an adjustable scale to take care of that. Since the airspeed indicator is connected to both the pitot tube and the static pressure port, any interference with either of those sources upsets its reliability.

Allied with the pitot-static system are two ancillary systems for protection and redundancy: the pitot heater, which prevents the formation of ice within the pitot tube; and an alternate static pressure source. The alternate source is pretty important, since it affects all three pitot-static instruments. It usually consists of a valve under the panel or somewhere accessible to the pilot. Since the static pressure readings taken from inside the cabin are not an ideal source for pressure measurement, you'll get an error in your instrument readings when you operate from the alternate source; most operator's manuals and airplane flight manuals give a correction factor for altitude and airspeed with the alternate source. I think the best way to learn how to apply those corrections is to operate the valve sometime when you're flying VFR and see how the instruments vary when you switch between primary and alternate. You'll learn firsthand how much and in what direction the readings change, and you'll probably never forget it, either.

An emergency alternate static pressure source can be had by breaking the glass on the face of the VSI. Obvious question: with what? Nobody ever tells you that. I've never had to break one, but I've thought about it and decided I'd use the back end of my two-cell flashlight. Hope it works, because I'm not about to carry a little hammer around.

Static ports are little holes in the sides of the airplanes and have no moving parts — the models of reliability. Still, people manage to take off with the ports clogged with insects, the results of the last wax job or paint. Pitot tubes also beckon to bugs like a motel with a vacancy, and many operators keep a pitot cover over the opening — only to depart with the cover still in place. (There is more than one way to wreck your day.) Visible blockage of the pitot-static system is easily detected through a preflight, but the most difficult problem to detect is the kind you can't see: water in the air lines. Water can collect at low spots through condensation or by rain finding its way in, and the dollop of water that finds its way in usually stays there at the lowest point. That's why many airplanes have static system drains built into their lines to allow you to vent the "sump" as you would a fuel system. Water in the pitot-static system can affect all three of its respective instruments by freezing at altitude or otherwise blocking the tube.

By virtue of its being located at the top center position of the standard panel arrangement sometimes known as the "basic T," the most prominent gyro instrument in most panels is the attitude indicator or artificial horizon, the former name being preferred these days. The value of the attitude indicator is that it provides an easy visual reference for two axes of motion — pitch and roll — in a single instrument; notice that this doesn't say it's the best visual reference, just easy. The horizon, or attitude indicator, is the instrument that comes closest to duplicating one element of the familiar outside world (the natural horizon line). Part of the discipline of learning to fly on instruments includes acquiring a taste for the other five faces surrounding the attitude indicator, because although the attitude indicator gives a very good general picture of pitch and roll, it is precise for neither.

The attitude indicator is based on a self-erecting gyro with its spin axis arranged vertically when the airplane is level. Applying the rule for gyroscopic forces, you can see immediately that this gyroscope will give you no information about the airplane's motion around this same vertical axis — in other words, no yaw information, a fact that you can demonstrate to your satisfaction while taxiing. The term self-erecting means that the gyroscope is attached to a mechanism that exerts forces to align the spin axis to the vertical automatically. This mechanism, whether part of an air- or electric-powered gyro, is usu-

The artificial horizon or attitude indicator as it would appear in a descending right turn. The index at the top — an inverted V within a clear plastic pointer — shows that we're in a 12-degree bank. The major divisions around the upper half of the instrument are marks for 30 degrees, 60 degrees, and 90 degrees of bank. The 10s and 20s in the middle of the instrument register degrees of pitch up or down. The converging lines simply help to enhance the display, although some people use them for bank reference. The word "vacuum" leaves little doubt that this instrument is powered by air, which is typical design for lightplanes. The adjustment knob at bottom center serves only to move the little airplane symbol up or down vertically to adjust for different eye levels.

ally very simple and operates by sensing the pull of gravity. If the gyroscope should wobble from its appointed orientation, the erecting mechanism compensates in some fashion, either by opening little air slots or by allowing some bearings to create a correcting counterbalance until the gyro realigns itself properly. As you can easily imagine, minor errors arise if lateral acceleration is applied to this instrument in such a way that a G-force pulls sideways on the erecting mechanism, but they're not bad enough to disorient you and they're only transitory.

Some older attitude instruments used to have a caging mechanism to capture and hold the gyro wheel in the proper position. Very few newer lightplanes provide a caging knob, offering only an adjustment to move the little airplane symbol up or down with respect to the horizon reference line.

The attitude indicator's primary role is to replace your sense of "down." Granted, it also tells you if the wings are level, which is derived directional information, but the attitude indicator tells you nothing at all about the compass direction in which you're flying. There are no landmarks on its little horizon and you could fly forever in its tiny world and discover nothing.

Directly beneath the attitude indicator is another universally mounted gyroscopic instrument that tells you what the attitude indicator cannot: direction of travel. The *directional gyro*, also called the heading indicator or simply the DG, spins around a horizontal axis that's in the same geometric plane as the wings of the airplane. This axis is also free to rotate, however, so that the gyroscope is capable of moving around a vertical axis and thereby driving a compass card display. Therefore, depending on how its rotational axis is oriented, the airplane in pitch maneuvers may create precession forces that lead to errors.

Directional gyros are still built with a caging mechanism that also serves as an adjustment and aligning device. This instrument cannot sense magnetic north, so you have to adjust it periodically to agree with the magnetic compass. The gyro is meant to duplicate the information from the magnetic compass without all the hassles the compass imposes. Magnetic compasses jiggle and whirl in flight to the point where they are practically useless for attitude flying; consider them a standby emergency instrument but no more. Also make sure that the magnetic compass is not in the midst of some tango when you use it to set the DG.

The directional gyro is valuable for its steady presentation of directional information, but it does have certain built-in errors. For some reason, the term precession has come to mean "error" when applied to DGs, although that usage is not strictly accurate. Directional gyros have to be aligned about once every 15 minutes or so. Although no regulation says so, a gyro that drifts in level flight more than 3 degrees in that 15-minute interval is generally considered unacceptable.

This directional gyro indicates a magnetic course of about 251 degrees and a reciprocal heading of 071 degrees. The index line off the tail of the little airplane provides a quick reference to obtain reciprocals, just as other markings — each wingtip has a smaller line just outside it, and at 45 degrees left and right, there are two circles—offer references for 45-and 90-degree course changes.

This instrument has to be checked against the magnetic compass periodically and adjusted by pushing in and turning the knob in the lower left-hand corner (and slightly out of focus in this closeup); the knob also serves to cage the gyro itself. A heading "bug," this one set for a course of 140 degrees, is just a manually adjustable reference set by the knob in the lower right-hand corner; it helps you to remember headings assigned by ATC. Although it doesn't say so anywhere on it, this DG is air-powered and has no power-loss warning flag.

This magnetic compass — sometimes called a wet compass because its card floats within a dampening fluid of white kerosene — is the primary heading information source on airplanes without a gyro compass that's slaved to some automatic source of heading information.

In anything but straight-and-level flight, however, this compass can't be used for attitude flying without great strain and practice. The primary source of directional guidance, then, is the directional gyro, which is adjusted to agree with the magnetic compass here every 15 minutes or so.

This compass shows a magnetic heading of about 312 degrees. The numbers on top and bottom of the window give corrections to account for differences that arise in this airplane. It's important to understand that such correction cards are calibrated with the airplane's radio systems and electrical systems turned on, and with everything turned off, the compass may read slightly differently.

Just above the compass is the outside air temperature indicator, which you use to obtain true airspeed data.

What actually happens in flight is that precession forces tend to move the gyro from its ideal rotational plane. A compensating system is incorporated into the design of the instrument, but in performing its realigning function, the heading card may drift. Also, if the gyro is aligned in such a way that a pitch change creates precession, a long flight such as one from Los Angeles to New York may create an error. Think about it this way: when you fly from L.A. to New York, you are quite literally diving the airplane around the surface of a sphere. Believe it or not, it's approximately a 45-degree nosedown pitch change, and that is more than enough to affect the gyro. The main force at work when the DG card wanders from its proper orientation to magnetic north is internal friction in the bearings, particularly in lightplanes that aren't capable of long, high-speed flights that result in rapid and large longitude changes. Still, it's good practice to cage the directional gyro periodically by operating the compass card adjustment knob, whether the DG needs aligning or not. The act of caging the DG restores the gyro's flywheel to its ideal orientation.

In that orientation, the gyro will always sense yaw motion. In coordinated flight, it is the roll axis of the airplane that produces turns and the rate of turn, and roll is produced by the ailerons, creating in the pilot's mind a functional relationship between the DG and the ailerons; nonetheless, the DG primarily senses yaw (and internally disposes of pitch and roll). To demonstrate this, taxi an airplane and watch the DG register changes of direction as you turn corners. With the wheels flat on the ground — the airplane *can't* roll now — the instrument performs perfectly. Still, when it comes time to fly, the DG is the most accurate and sensitive indicator of roll motion and bank angle in the panel. How can this be, since the instrument simply casts aside the effects of roll?

In a flat turn similar to the situation when you were taxiing the airplane around a corner, the DG delivers pure yaw motion. Trouble is, there is no normal maneuver in ordinary attitude flying (unless you want to argue that flying the ILS with the rudder pedals qualifies) that corresponds to that form of motion. The apparent perceptual conflict that's embodied in the DG is that it tells you about change in lateral direction, and the only way to produce such a change in coordinated flight is to move the ailerons and bank the airplane around the *roll* axis. Although the instrument seems to tell you more about roll and bank, it actually senses yaw. Don't be too hard on your instructor, therefore, if he or she tells you that the DG is the primary instrument for roll or words to that effect. What the instructor is really saying is that the DG is so sensitive to the effect of slight bank angles that it is your most dependable and precise long-term directional cue in coordinated straight-and-level attitude instrument flying. Your primary directional chore in that situation is to maintain your heading, and even imperceptible bank errors will show up on the DG most clearly.

DGs lend themselves to one human failing: leaving them caged. A spring return on the adjustment and caging knob has cured much of the problem, but many instructors still like to see you double-check for uncaging by giving the knob a little twist after it's been used to adjust the card.

Almost all new directional gyros depict an airplane with its nose pointed up toward the top of the panel and with a vertical compass card that rotates around this nose index or "lubber line." This design has the advantage of depicting motion that resembles the actual direction of the turn. If you turn to the right, the airplane symbol appears to swing to the right against the card's motion.

There is another, older form of directional gyro you may encounter, though, and it must be treated with a degree of caution. The so-called barrel type presents you with a strip of numbers that move behind the lubber line in the same way that the magnetic compass moves in its little bath of juice. The problem with this presentation is that the pilot experiences a sensation that looks as if the opposite wall being viewed through the window of the instrument is moving the wrong way. The instrument's worst characteristic is that it can induce a sensation of a turn in the direction *opposite* what you intend. In order to beat that feeling, you must discipline yourself to remember that all turns to the right cause the value of the compass headings to rise. All turns to the left cause them to drop. That's easy enough to say, but getting in there and doing it is something else again. I hate barrel type DGs. Don't know anyone who likes them. That's why they've largely disappeared. Throw yours out and buy a new one.

So far, we have two instruments that tell us a lot about the airplane. The attitude indicator at the top center gives us a good general picture of the situation in pitch and roll, and directly beneath it, the directional gyro, by sensing changes in yaw, gives us precise indications of the roll situation. Let's pause right here and assess the airplane and the instruments in level cruising flight. It's true of most airplanes in this configuration that they are stablest in yaw, somewhat less stable in pitch, and least stable in roll. What does this tell you about the chore so far? It appears that you'll be spending an enormous proportion of your piloting workload on keeping the wings level. (If you don't believe that, try flying an airplane with a simple wing leveler. Turn it on for a while, then turn it off. Which keeps you busier?) The arrangement of attitude indicator with the directional gyro directly beneath it has evolved empirically as a result of the nature of the flying task. Those two instruments command the center of your attention.

Another instrument that provides directional information is located at the lower left in the basic group of six, and it was the first of the gyro type to find wide application in instrument panels. The *turn indicator* and its accompanying inclinometer (they're two separate instruments that have over the years simply been incorporated into a single case) are also called the "needle and ball," the "turn and bank," and most recently, the "turn coordinator," which is really a very different instrument from the older needle type.

To make a turn indicator, you simply attach a spring to the same sort of gyro you set up for the directional gyro. Instead of turning a card, though, the precession force of the gyroscope in a turn pulls against a spring with a force proportional to the rate of turn. A mechanical linkage inside the case reverses the motion of the gyro with respect to the needle so that the needle points in the direction of the turn. The ball of the inclinometer is there to allow you to

The electrically powered turn indicator senses the rate of a turn. This newer type of indicator is called a turn coordinator because its gyro senses some roll as well as yaw. A standard rate turn is one in which the wingtips of the airplane symbol align with the indexes marked L and R; the "2 MIN" marking refers to the time it takes to turn 360 degrees at a standard rate using this instrument. The warning flag for loss of electrical power can be seen just above the right wing: if power fails, a little red panel fills the window.

The inclinometer, or "ball," is just below the word "turn indicator," and this one works in dampening fluid. Right now, it looks as if we need a touch of right rudder.

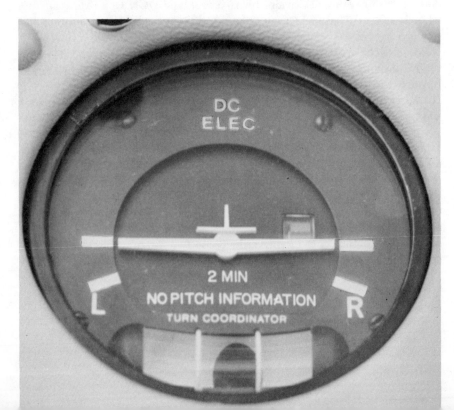

coordinate rudder action and trim the airplane for yaw. As such, it is the only instrument on the panel with a functional link to the rudder, the control surface that provides yaw maneuvering. The ball is free to roll back and forth within a curved glass tube filled with a fluid to dampen the ball's movement; some instruments use a mechanical ball symbol that's dampened internally. If the damping fluid is missing, the inclinometer will do nothing but alarm you and is relatively useless.

The ball functions as a sort of pointer, indicating the location of the gravity vector of an airplane in any situation. Coordinated flight requires that the gravity vector remain centered on the vertical axis of the airplane — even if the airplane is in a steep bank (see figure, page 00). In order to accomplish this, the centrifugal force generated in the turn must be balanced by tilting the lift vector toward the inside of the turn just enough to counterbalance the centrifugal force. The ball signals any imbalance between the inward-directed lift component and the centrifugal force. In an emergency, pilots have used a liquid in a container — a cup of coffee, for example — to substitute for the ball. If the liquid level skews to one side, you apply rudder pedal on the corresponding side. The inclinometer is a very simple instrument, and an important one too.

Rate of turn is not a very important piece of data for IFR flight unless you pride yourself on flying precisely. As a student, you'll be expected to execute turns at the *standard rate* or half standard rate, depending on the situation. A standard rate turn is one that changes heading at 3 degrees per second; half standard rate is 1.5 degrees per second. On the instrument, the manufacturer provides an index mark for zero rate and two alignment indexes for standard rate turns in either direction. On the older needle type, the symbol is a little house, usually referred to as the "doghouse," although it reminds me more of the architecture in those little houses they give you with Monopoly games. Turn coordinators provide a little airplane symbol and two marks with which to align the wingtip for a standard rate turn.

The trouble with the old-style turn indicators was that they presented pure yaw information, and only yaw information, including adverse yaw. Thus, the first dab of down aileron and its resultant drag-induced adverse yaw would present an indication of a turn in the wrong direction, at least momentarily. To cancel that effect, an instrument called a turn coordinator has been developed. By tilting the gyroscope 30 degrees from the horizontal-axis orientation, the gyro also becomes slightly sensitive in roll, just enough so that the adverse yaw characteristic is eliminated and the instrument can also provide a fairly good wing-leveler display. Bear in mind that because of its design, the old turn-and-bank really was awful as a wing-leveler indicator, yet we still think of it that way and many pilots associate it with roll control. The new turn coordinator is a much better backup instrument (also, it's electrically powered in the majority of cases) in the event of loss of the other gyros than the older type used to be. For one thing, rate of turn relates to bank angle only after you have considered the airspeed. At high airspeeds, you must bank the airplane much more in order to obtain the same rate of turn that you got at a lower airspeed with less bank. Thus, the needle points in the direction of the turn but provides only indirect information about angle of bank and rate of roll. To demonstrate that the instrument really senses yaw rather than roll, observe its display while taxiing around a corner. Even the newer turn coordinator, which is more talented at roll sensing, behaves strangely in a taxi turn. Turn left on the taxiway and the airplane symbol will *bank* left, even though it's obvious the airplane cannot bank. Therefore, although the turn coordinator is better than the old needle, neither instrument is entirely accurate with respect to the roll axis, the needle less so. Do not fall into the trap of allowing display symbology to distract you from the knowledge that the instrument displays rate of turn by sensing yaw. The turn coordinator senses some roll, but it is far more sensitive in yaw.

The gyro instruments are a great source of confusion because you must gather information from all of them in order to gain an accurate picture of the airplane's performance around the roll axis. We already agree that the attitude indicator gives the best general picture of the wingtips with respect to the horizon, but any proper turn requires that we check that picture against the turn coordinator for its depiction of rate. Until we have flown an airplane long enough to know which bank angles correspond to standard rate turns at various airspeeds, we cannot know the rate of turn from the attitude indicator alone. Why? Because airspeed influences the rate, and that makes the turn coordinator the only absolute authority on rate information. Similarly, the only source of directional information in a turn is the DG. A turn has to end somewhere, usually on some heading reference, since that's the way ATC works. Without a DG, you are "no gyro," and you will operate under a different regimen, which we'll talk about under partial panels and other unusual procedures.

So roll and yaw axes are frequently very confusing, since a turning maneuver is a combination of both. Maybe pitch change will prove to be easier to figure out. We already have one gyro giving us a pretty good indication of pitch — the attitude indicator. The only lack in the AI is genuine accuracy, some measure of what the pitch configuration is actually doing for us.

Pitch has this incestuous relationship with power. It is perfectly reasonable to expect the airplane to descend while its nose is pointed above the horizon reference, so there is no direct relationship between where you point the airplane and where it will ultimately go unless you also consider power. Therefore, pitch information alone is of little use, and that muddies the role of the attitude indicator right at the outset. At cruise power, which is how you'll spend most of your IFR flight, any movement of the elevator that results in a pitch change will register immediately on the attitude indicator. Thus, the AI is an excellent instrument for finding a reference attitude during transition from level flight to a climb or a descent. Once you've established the attitude, however, you need more information about what the change in pitch has done for you, and the best and most sensitive instrument for that, after you've used the AI, is the altimeter.

The altimeter is the first of the pitot-static instruments we'll consider, and the reason why it's so important to pitch control is that very minor changes in pitch, even so small as slight out-of-trim conditions, register very plainly and rather quickly on the altimeter. There is some lag before the instrument responds, but the scale of the altimeter in units of 20 feet enables you to maintain the aircraft in its pitch axis with a high degree of accuracy as measured by its vertical position in the air.

The airspeed indicator is also a very good source of pitch information if you're willing to accept its drawbacks. Airspeed is primarily a trim function, for one thing, and you can push the nose up or down through quite large excursions of movement through the pitch axis and it takes a while for the speed of the airplane to reflect the new pitch attitude. In cruise, if you leave the power constant but adjust the pitch trim, the airspeed indicator will slowly register the change. If you leave the trim constant and alter the power setting, a well-rigged airplane may remain glued to the airspeed but change its pitch attitude and altitude rather briskly. Thus, the airspeed indicator is as closely related to power management as it is to pitch trim and control.

Because all three instruments — the attitude indicator, airspeed indicator, and altimeter — relate to pitch in slightly different ways, it seems logical for them to be arrayed in a single row across the top of the basic T.

Now all you're missing is a rate instrument for pitch that tells you how fast things are changing as the rate-of-turn instrument did for bank/roll/yaw. The vertical speed indicator, or VSI, the one with the leaky aneroid, takes care of vertical rate data by presenting the pilot with a needle that points to hundreds of feet per minute up or down. In general, VSIs are prone to lag, but the instantaneous VSIs are an improvement and can be relied upon for a valid reading within seconds of pitch adjustments; some of them respond earlier than do altimeters.

The altimeter also provides rate information, in a way; if the needle moves fast, the altitude is changing quickly. The VSI is nice to have for longer climbs and descents, however, when you need to know your climb gradient. An obstacle that must be cleared during a climb may require that you figure out in advance whether you're going to make it or not. The VSI can answer that. Most changes in altitude are assumed by ATC to be accomplished at a rate of about 1,000 fpm until you reach a level 1,000 feet from that assigned to you, at which point you slow the rate to 500 fpm until finally leveling off; the VSI is a big help here too.

Some methods of instruction may put more emphasis on using the vertical

Two forms of altimeter, one with a pointer for hundreds of feet and digital numbers for thousands of feet (above). The warning flag—a striped bar across the thousands numbers—indicates that electrical power is off to the altitude reporting feature, which attaches altitude readout to your radar transponder reply to the ground. This altimeter shows 2,090 feet. Like all altimeters, it must be set to a reference barometer by means of the adjustment knob in the lower left-hand corner. Just under the word "electric," you can see the barometric setting, which right now is about 29.915.

The nonreporting altimeter (below), using clock hands and without the digital display of the other, is the type you'll meet in most training aircraft. The very long hand indicates hundreds; the shorter, fatter hand indicates thousands; and just behind the long hand is hidden a very small hand that indicates ten thousands of feet. This altimeter reads just over 2,100 feet, and it is set to a barometric standard of 29.97 inches.

An airspeed indicator calibrated in knots, which is the unit of airspeed that air traffic control people use, making this a good instrument for IFR. This one also has a sliding scale to correct for nonstandard temperatures at cruise altitude. You align the temperature scale with the pressure altitude scale and the sliding ring moves that outer series of numbers around so that the pointer indicates true airspeed instead of indicated, thereby compensating for your altitude and the nonstandard temperature. If you insist on talking in miles per hour, they've built a little window into the center of the instrument that reads mph directly. This airspeed indicator belongs to a Cessna Centurion, and as you can see, there's a line just under 200 knots called the redline and above which thou shalt not fly. Other operational arcs range from the redline down through the flaps-down stall speed, at about 55 knots. An airspeed indicator can therefore tell you at a glance a number of key operating data about the airplane.

This vertical speed indicator is calibrated in hundreds of feet per minute through a total of 1,000 feet per minute, but reads 1,500 feet per minute and 2,000 feet per minute. It requires no adjustment and isn't even a required IFR instrument. Few airplanes leave a factory without one, though. The VSI is an emergency static pressure source: you break its glass face and allow ambient air to get in. Obviously, the instument itself then becomes useless. Most serious instrument-equipped airplanes now offer an alternate static valve buried somewhere under the panel.

speed indicator for pitch cues as an adjunct to the attitude indicator, and to some degree, it depends on the particular instrument's behavior. It's true, though, that some VSIs are very good at providing a nearly immediate indication of the *trend* produced by a particular attitude. The VSI needle is really good for telling you two things: "You are going up" or "You are going down." If you want to know how *fast* you're doing either, you have to wait awhile for the instrument to stabilize. Still, it is a good, sound method to add the VSI to your scan after beginning a climb or a descent to assure you, through the pitot-static system's independent sensing, that the gyro attitude indicator is telling you the truth. Practice with your altimeter and VSI to observe how they respond to pitch changes, then use both as appropriate to give you a pitch attitude backup.

Once you know how all the instruments work, you'll be better able to put them to use in actual attitude instrument flying, and the fundamental basis for successful panel work is the correct use of each instrument at the appropriate moment. Unfortunately, someone once labeled the task of instrument monitoring with a term that has stuck — "scan." The word "scan" implies to entirely too many pilots a kind of random jumping from instrument to instrument, and they almost feel guilty if they haven't given, say, the VSI its quota of attention. That's nonsense. Nobody can tell you in what order to scan the panel or how much time to spend on each instrument.

Successful attitude instrument flying, a necessary precondition to navigating IFR, is simply a matter of connecting maneuvers to functional responses from the pertinent group of instruments. Straight-and-level flying is pretty easy, descending turns are a little harder, but all instrument maneuvers are a matter of learning to manipulate the controls in a smooth and correct manner, using the instruments to tell you moment by moment exactly how your control inputs match your expected result.

Our model of perfection, the flight director, is able to do it by integrating mechanically all the data from the instrument sensors. As we pointed out in Chapter I, it does this by adapting to varying rates of change and matching the control command to the situation.

For IFR, there are five pitch situations that pertain: climb, cruise, en route descent, approach, and approach descent. Those, combined with random combinations of turning maneuvers, form the basis for every IFR flight. Notice that although an IFR flight has no predictable pattern of turning maneuvers, the changes in pitch adhere to the above pattern fairly strictly: on every flight, you will climb, cruise, and descend. One common denominator for all turns, though, is that they are supposed to be accomplished at standard rate. So a good place to start taking apart attitude instrument flying is with turns from level flight.

Turns are accomplished by banking the airplane to an angle sufficient to provide a standard rate, maintaining altitude with slightly increased back pressure. A banked turn increases the apparent weight vector of the airplane and induces greater drag as the lift of the wings counterbalances the weight. If you don't like the term "weight," then use "G"; it's the same thing. Finally, the entire maneuver gets unwound out of itself in a return to level flight. Leading your desired heading by a time sufficient to provide for a smooth maneuver, relax back pressure as the airplane is rolled back to wings level with the ailerons.

What's changing? Bank angle, pitch trim, and airspeed (due to increased drag). In order to produce the proper bank angle, we'll first watch the attitude indicator as we roll into the turn, then glance at the turn coordinator for a cue as to when to halt the roll. For most lightplanes, about 15 to 17 degrees of bank does the trick, and after we've acquired the proper angle, we can afford to spend less time with the rate-indicating turn coordinator.

Having established a bank angle that will provide a standard rate turn, we turn our attention to the change in pitch force to maintain altitude. Again, the attitude indicator gives us a good general appraisal of the slight nose-up adjustment we'll need, followed by attention to the altimeter for fine corrections. The VSI may help by offering a peripheral trend cue, but the altimeter should get our major attention. Last, since it lags somewhat, we check airspeed, particularly if we're turning at a low airspeed, so that we can add power if necessary.

Unwinding into the roll-out, we watch the directional gyro, alternating between it and the attitude indicator, occasionally checking the turn coordinator to ensure that the rate is dead on. A typical standard rate turn might call for 15 to 17 degrees of bank, so at a heading 8 degrees from our desired roll-out heading (following the rule of thumb that says "lead by half the bank angle"), we use the attitude indicator to begin a smooth roll-out to wings level. Practice to determine your own roll-out data, and also learn to anticipate the relaxation of back pressure so that you don't have to glance at the altimeter or VSI for altitude corrections as you level off.

We have produced a maneuver with one gross manipulation — a 15- to 17-degree bank — and one very minor manipulation — the pitch pressure change. The *rates* of motion about the two control axes (forget about yaw unless you can't keep the ball centered, in which case, you should go back to VFR for a while until you can) are entirely different and therefore call for different periods of attention to different instruments as their particular talents apply to the situation. It would make no sense at all to fixate on the turn coordinator once you've established a standard rate turn. You watch it only long enough to tell you when the bank angle on the AI is just right, and once you get to know that, you may not even check it at all!

The pitch correction is a very tiny adjustment compared to the bank, requiring no gross movement of the elevator and therefore no undue attention to the attitude indicator or altimeter. You can feel comfortable in dropping down to the DG now and then to check the progress of the turn and anticipate the roll-out lead.

I know pilots who trouble themselves with mental timing calculations during turns, but to me, it's just confusing and too much distraction. What really ought to be going on here is a very simple task of orderly review of instrument faces that match the task at hand. The flight director takes care of the job in an unruffled way: command roll to a bank angle that provides standard rate; slow rate of roll for en route turns, faster for approach corrections; smooth, well-damped commands for en route, "peaky" commands for approaches; pitch-correction commands as required by altimeter excursions, with very minor displacement of the command bars at cruise airspeeds, somewhat larger displacement at approach airspeeds, initially higher rate of roll command on the roll-out, tapering to a lower and lower rate to wings level.

Once you feel comfortable with turns in level flight, try the maneuvers that call for pitch changes: climbs and descents, then constant-altitude speed reductions. The order doesn't matter. Begin by watching what happens when you make a change in pitch attitude. First, try a climb without touching the power: attitude indicator first — about a bar's width above the horizon line is a good place for starters — then watch the VSI and altimeter for their first response as they begin to register a gain in altitude, then the airspeed indicator as it begins its creeping to a lower airspeed. Notice that the attitude indicator's rate of response is nearly immediate, while the altimeter, VSI, and airspeed indicator are slower to respond. Give the attitude indicator your major share of attention as airspeed slows. Notice that if you simply move the elevator once and keep it there, you cannot maintain the bar's width above the horizon. Why? Because as the airspeed reduces, flow over the elevator decreases and reduces its pitch-trim authority. If you're going to keep that bar's width reference, you'll have to add some elevator back pressure — or retrim. This sequence of constant cross-checking between the attitude indicator and the airspeed indicator continues until the airspeed settles on a stable value.

So far, we've ignored the altimeter except as an initial reassurance that we were indeed climbing. Now add the requirement to climb to a specific altitude. The attitude indicator is your primary cue as you transition into the climb, and it later becomes important once again as you strive to maintain the climb attitude you want.

We happen to know, however, that airplanes all have ideal climb speeds, and we really ought to adjust that attitude reference over the long term to match the ideal airspeed for best rate of climb.

Duck soup.

Airspeed changes very slowly, right? Use attitude initially as your primary reference, because the airspeed will change more rapidly at first, then begin to slow. If you watch closely for the point at which it begins to slow, it will give you sufficient advance warning of where it's likely to *stop* changing so that you can adjust the attitude promptly and smoothly. Now the airspeed change has smoothed, tapered in rate so that the airplane will approach its ideal climb airspeed gradually, just the way a flight director would achieve it.

The attitude indicator and the airspeed indicator, in conjunction, deliver the pitch information you need to establish the climb. The VSI is irrelevant except in cases where it might exceed 1,000 fpm — some airplane! — or where you wish to monitor the rate over the long term to ensure obstacle clearance in a climb. Even though the altimeter is supposed to be our most accurate instrument for pitch information, it sure isn't telling us anything relevant right now. We're looking for an airspeed. And in looking for it, we could very properly be accused by conventional flying wisdom of violating the rule that prohibits *fixation*. Aside from occasional glances at the inclinometer to assure us of yaw trim for propeller effects in the climb, we've spent the entire time staring at the AI and the airspeed!

Only one ingredient missing: we've got pitch all right, but in order to climb at the ideal airspeed, we'll need to add power to that. Big deal. One way to handle power changes is to establish the climb attitude first, allow airspeed to drop, then add power. Remember to move the prop handle before adding manifold pressure on the so-called complex airplanes, but you already know that from VFR flying. Some airplanes climb at full throttle, others call for a limited manifold pressure setting, commonly 25 inches for normally aspirated airplanes. As the airplane gains altitude, occasional glances at the altimeter — and I mean *occasional* — enable you to anticipate the level-off with ease.

Descents are slightly different. Let's try the same exercise as last time, leaving the power alone for a moment (although it would be wise to set the power fairly low for this maneuver) and just pushing the nose down to see what happens. Checking the attitude indicator first, establish a bar-width nose-down attitude by adding forward pressure to the stick. The altimeter will quickly register a descent as the needle starts to unwind, and the airspeed indicator tells you you're picking up airspeed. Hold the same down *pressure* on the stick, and what happens? The nose begins to rise! Amazing! You find yourself doing a similar but reverse manipulation from what you did during the climb: you have to add more control displacement to hold the bar-width nose-down attitude. The reason for the change is similar too: airspeed over the tail has increased, thereby altering the authority of the restoring force of the trim surfaces on the tail. In order to maintain the bar-width nose-down attitude, you had to add control *pressure* (or call it trim) in pitch when power remained the same. Because the nose tends to rise in this sort of a descent, instrument pilots learn that the pitch maneuver to descend an airplane is initiated by reducing the power some practiced amount, about 5 inches of mp or 500 rpm being typical for light singles. This practice is known as "power trimming," and with it, you can establish very smooth transitions from level flight to a descent without fussing with trim pressure changes. You'll also find that in almost all airplanes, a simple power reduction just naturally leads to a nose-down of about a bar width, maintaining a painless cruise airspeed, with a 500-fpm descent.

For this maneuver, the first instrument you'll watch will be the power monitor (tach or mp gauge) for initial setting, followed by attitude indicator for some reasonable suggestion that the nose has dropped. After that, check the altimeter, which will have begun to unwind, the airspeed to ensure that it's not changing radically, and finally, the vertical speed indicator for your rate of descent.

Since the maneuver was initiated with a power change and very little elevator manipulation or pitch-trim change, it makes sense to continue to use the power to influence the rates you're most interested in: airspeed and rate of descent *Once the descent is established by the power reduction*, you will make very minor pitch corrections with elevator and trim in order to keep the attitude, airspeed, and descent rate within bounds.

Consider the case again if you adhere to the philosophy of very high airspeed descents, maintaining cruise power while you lower the nose. As we've already agreed, you'll have to alter the pitch-trim pressure to compensate, pumping in more nose-down trim. What usually follows an en route descent? Slowing to approach airspeed. Why speed up only to slow down again? You're imposing on yourself some very distracting extra chores in changing the pitch trim when you could accomplish the descent and the ensuing level-off at your lower assigned altitude using the power alone, followed by, at most, a *single* minor pitch-trim change when you slow to approach airspeed and drop flaps. You'll also be making gradual power reductions in steps rather than a single, drastic power reduction with rapid thermal cooling of the engine metal.

To summarize: for IFR climbs, set attitude, wait for airspeed, and add power; for IFR descents, reduce power, allow the nose to drop to a reference attitude (a bar's width is nice), monitor airspeed and rate of descent. Adjust airspeed with pitch trim and rate of descent with power.

Instrument monitoring concentrates primarily on the attitude indicator and airspeed, in both situations, with occasional glances at the altimeter to see how you're doing and at the VSI to monitor the rate. Since climbs occur at full power or some climb power setting, you simply accept the best rate upward you can get. In a descent, the VSI has an enhanced role in providing cues for power adjustment.

You have just gone the flight director one better, because it gets no power information at all! If you were to ask a flight director to hold a specific airspeed in a descent, it would do so by changing pitch trim and moving the nose up and down to adjust. You can achieve a much smoother descent by using power to adjust for the more important area of performance: vertical rate. Notice, though, that flight directors level off from descents very gradually, with plenty of lead. They do this to minimize structural loading on the wings, and you should strive for the same smoothness in reestablishing level flight after a descent. Here too, "power trimming" will help you by keeping airspeeds very close to cruise airspeed in the descent.

To review the instruments in terms of their sensitivity to changing rates:

The attitude indicator responds quickly and displays a general picture of roll and pitch. For very precise measurements of both, you need to add —

the directional gyro, for roll, and the altimeter, for pitch, mostly during straight and level flight, and also —

the turn coordinator, for establishing the standard rate turn in conjunction with the attitude indicator, plus —

the airspeed indicator and VSI, in conjunction with the attitude indicator, during climbs and descents.

People make airspeed reductions into overly complicated maneuvers, when all they really are is a variation on the climb with a reduction of power, and with extra attention to the altimeter for pitch correction cues, since you want to hold altitude extra precisely. When you're *changing* altitude, the altimeter actually decreases in importance. Sure, you check it now and then for your progress, but you watch it much more closely when it's important to hold precisely the same altitude during any maneuver.

The same is true of the directional gyro while you're *changing* direction in a turn. Just as with the altimeter during an altitude change, you glance at the DG during the turn to check on your progress toward your new heading, but most of your attention is spent on the attitude indicator, altimeter, and turn coordinator, generally in that order of importance.

Applying our model flight director and our consciousness of rates of change to both situations — altitude change and turns to a new heading — we observe that both the altimeter and the DG can give us some idea of rate simply by our watching how fast they're changing. But the two instruments that deliver *pure rate*, the turn coordinator and the vertical speed indicator, gain a great deal of importance over their roles in straight and level flight.

I hope that by now you can begin to see that an instrument "scan" is not random at all but a carefully considered choice of groups of instruments whose capabilities match particular situations. Whereas visual flying asked only that you keep the wings at a semblance of level and the nose somewhere around the horizon while you're straight and level, IFR flying demands more: use of the DG to hold a precise course (the IFR equivalent of wings level) and the altimeter to maintain precise pitch (the analogue of the nose reference visually).

Part of practicing IFR attitude instrument flying is the gradual learning of each maneuver, then the practice of one maneuver followed by another, and finally, combinations of altitude changes and turns executed simultaneously. The key to success, though, is a thorough understanding of each instrument and the situation in which it helps you the most. Once you've acquired that knowledge, you'll be way ahead when it comes time to practice under the hood or in a ground trainer. Sorry, but knowledge alone won't make you a good attitude instrument pilot; that takes practice and eye-hand coordination. That's right — you go out and *do* it.

IFR Notebook

Equipment required for IFR flight:

1. Basic day VFR equipment, which includes —
   airspeed indicator
   altimeter
   magnetic direction indicator (a compass)
   tachometer for each engine
   oil pressure gauge for each engine
   temperature gauge for each engine
   manifold pressure gauge for each engine
   fuel gauge
   landing gear position indicator (retractables)
   seat belts and shoulder harnesses
2. For night flight, add the following —
   position lights
   an approved anticollision light
   landing light, if for hire
   adequate source of electrical energy for installed equipment
   one spare set of fuses or three spare fuses of each rating required
3. For IFR flight, add the following to the day VFR list or the combined day VFR and night VFR lists for night operation —
   two-way radio communication system and navigation equipment appropriate to the ground facility to be used
   rate-of-turn indicator
   slip-skid indicator (the "ball")
   adjustable altimeter
   a clock displaying hours, minutes, and seconds (either hands or digital displays are acceptable)

a generator of adequate capacity

bank-and-pitch indicator (also called an attitude indicator or gyro horizon)

direction indicator (also called a directional gyro or DG)

4. For flight above 12,500 feet —

a transponder that meets the correct *technical standard order* (TSO C74c) and provides 4,096 codes, and also

an altitude encoder, unless you fly at or below 2,500 feet *above ground level* (AGL)

5. For flight above (but not including) 12,500 feet to and including 14,000 feet, for any part of the flight at that altitude that lasts more than 30 minutes —

oxygen for the pilot and required crew, who must use it

6. For flight above (but not including) 14,000 feet of any duration — oxygen for the pilot and required crew, who must use it

7. For flight above (but not including) 15,000 feet —

the operator must provide oxygen for each occupant of the aircraft, but for passengers, use is not mandatory

8. For flight at and above 24,000 feet in which VOR stations on the ground will be required for navigation —

an approved DME

9. For any flight (with certain limited exceptions) —

an approved emergency locator transmitter equipped with batteries that meet current requirements —

a) less than one hour's cumulative transmitting time

and b) newer than 50 percent of rated service life

(NOTE: The expiration is determined by the manufacturer and should be marked legibly on the transmitter case and entered in the aircraft maintenance record.)

About these rules:

The requirement for individual instruments for each engine sometimes causes confusion on twins equipped with a combined tachometer display in which both needles operate within a single dial face. Such a display is perfectly legal, since the term "tachometer" refers to the sending unit on the engine, not the display format.

Shoulder harnesses are actually required only on airplanes manufactured since July 18, 1978, and if you wish to seek refuge in that loophole, go ahead. After December 4, 1980 (which is a Thursday, in case you're wondering), and no, I don't know why, you'll need an approved metal-to-metal latching type. The rules require such a belt for every occupant older than two years.

That's right, you *don't* need a landing light to fly at night under FAR Part 91. Sometimes that rule can make quite a difference in conditions where a landing light actually reduces visibility.

Since most airplanes are now built with circuit breakers instead of fuses, the fuse requirement for a spare set aboard at night is outmoded. If you use an older airplane that has fuses for its circuit protection, it's a good idea to carry them as part of your flight kit in addition to any stored aboard the aircraft. This is particularly true for airplanes used by more than one pilot. Fuses are cheap and tiny.

Notice that there is a difference between night VFR and day IFR when it comes to sources of voltage for electrical systems (other than the magnetos). A battery is considered an "adequate source" of electrical energy for night VFR. A generator is specifically required for IFR flight; alternators, being merely another form of generator, qualify. Unless the airplane is placarded to cover exceptions, a full electrical load — the current with everything turned on — should not exceed 80 percent of the generator's rated capacity.

The phrase "appropriate to the ground facility to be used" means that you must figure out what ground equipment you'll be navigating with. A flight filed and cleared that involves only radar vectors direct to an ADF approach would not — hard to believe but true — require an omni receiver aboard the aircraft.

Notice, for example, that to fly IFR, a transponder is not required. The transponder requirements appear in certain regulations governing airspace but not those listing required equipment for IFR flight. Why the difference? IFR rules really relate to visibility, not to control of airspace. In controlled airspace, the rules may call for flight under IFR, but not necessarily. Controlled airspace is designed for air traffice control. Don't get IFR confused with air traffic control (ATC). They are two different considerations.

Notice that a vertical speed indicator is not a requirement in an IFR panel. It is perfectly legal to fly without one, or with an instrument that is malfunctioning, under IFR.

The rules stipulate that the rate-of-turn, bank-and-pitch, and direction indicators must be "gyroscopic," which is exactly what they've been for as long as anyone can remember, so don't worry about it. On large airplanes and helicopters, though, the requirement for the attitude indicator is slightly different from the rule for lightplanes.

In addition to the requirement that a transponder meet technical standard order C74c if it is to be used in controlled airspace (which is not the same thing as requiring *use* of a transponder in that airspace), there is a requirement for an encoding altimeter above 12,500 feet, except for in the airspace within 2,500 feet of the ground — which means mountainous areas, mainly. Since the encoder operates through the transponder on what is called Mode C, this is actually a requirement for a transponder as well. In case you are curious, the number of codes — 4,096 — is not happenstance but the mathematical result of transponders' using the octal code, with numbers zero through seven (count 'em; that's eight numerals). Octal codes are easily handled by digital computers of the type used in ground-based radar interrogation and ATC display systems. There are four windows on such a transponder. Multiply eight times eight times eight times eight and you get 4,096, or 8⁴ — the number of different codes between 0000 and 7777 you could possibly dial. The reason some codes, particularly the emergency codes like 7700, end in two zeros is because older transponders could select only the first two digits. They were called 64-code transponders and they're no longer acceptable, except for VFR use outside controlled airspace for radar assistance, which means they're practically useless. Transponders, incidentally, are also considered communication equipment in a pinch.

The requirement for oxygen is not limited to instrument flying but for any type of flight at the altitudes mentioned in the rules. Since required equipment seems to accumulate with increasing altitude, it helps to have all the requirements listed in one place. Pressurized airplanes require oxygen only when the cabin pressure altitude rises to the values listed. There are also additional requirements for pressurized airplanes operating at very high altitudes.

The DME requirement at 24,000 feet seems strangely worded. They probably meant to use the term "VORTAC," since a VOR station without TACAN or DME equipment paired with it can't respond to interrogation from a DME. At that altitude, navigation is by VORTAC anyway, since most of the high-powered ground stations have the DME capability of a VORTAC. According to this regulation, though, if you were navigating by some means that made no use of VORs, you wouldn't need a DME. It's all academic, though, because very few airplanes that operate at that altitude would come equipped without a DME. If the DME fails while you're up there, by the way, you have to report it to ATC, although they'll let you continue to your destination and then to an airport where it can be repaired (if there's no repair station at the destination airport). As soon as you descend below 24,000 feet, the rule no longer applies anyway.

About transponders and airspace :
Transponders, in addition to that sneaky requirement for their use above 12,500 feet that really rides on the coattails of the requirement for an altitude

encoder, are also required in Group I Terminal Control Areas (TCAs), as are encoders. The controller at the ATC facility in which you plan to operate can allow you to fly without a transponder if you make your request at least four hours prior to the flight. The controller can also waive the requirement for an encoder in any airspace, including the TCAs, if the transponder operates. Usually, the ATC people are pretty cooperative about this unless traffic is very heavy. The controller can also authorize flight without a transponder if the box has just gone belly-up in the middle of the flight and you want to land somewhere and get it fixed.

Encoders have their own special regulation, FAR 91.36, which stipulates that if ATC directs you to, you must switch off the Mode C, which is the switch position that enables the encoder to report altitude information along with the transponder code signals. The rule also requires that 95 percent of the time it is operating, the equipment must be within 125 feet of telling the same story as your altimeter, when the altimeter is set to 29.92. This requirement applies at the time of installation only, though. Frequently, you'll hear ATC tell a pilot that the encoder is reporting some altitude other than the one to which the aircraft has been assigned. The reason is because the controllers are allowed to use the displayed altitude on their radar screens (after an initial confirmation that it's correct) for traffic separation if the encoder reports agree within 300 feet with the altimeter readout to the pilot. If yours is off by more than 300 feet, it affects the way they handle you. Although the regulation may not require it, you should get the encoder calibrated to correspond to your actual altitude. Don't assume it's the encoder's fault, though; the altimeter may be in error.

CHAPTER V

Navigation, the "En Route" Procedure

This could be the briefest description of navigation you may ever read:

Navigation = attitude flying + wind correction

Now, to explain what that means: all navigators, regardless of whether they're piloting a ship or an airplane, face the same problem — to follow a planned path from one point to another. Even VFR flying involves point-to-point travel. Under IFR, you learn to fly an airplane solely by reference to attitude instruments and to control the airplane very precisely in its self-contained world. Getting from one place to another in the IFR cockpit adds the overlying problem of describing yet another system of references that define a *course*. Since attitude flying takes a fair amount of attention, it would be nice if the course-reference system could integrate handily with the panel instruments.

Attitude flying is the moment-to-moment chore of instrument flying, while navigation is the long-term task. Attitude flying involves relatively large and rapid changes that require immediate solution, whereas your airplane's progress with reference to its course and destination can appear almost painfully slow by comparison. Navigation is no fast-paced activity; you take care of it as you find the time between attitude flying chores. Your instructor may tell you "Aviate and *then* navigate." Same thing.

What gives new instrument pilots fits is that the navigation instruments occasionally behave as if things were happening very quickly. Therefore, part of learning to navigate on instruments under IFR is to acquire the ability to interpret your real situation in light of the rates of change that are truly important. "Chasing the needles" is one way of expressing the wrong way of going about it; we're hoping that you'll be learning it the right way.

It should be no surprise, then, that the methods for navigating are essentially identical, whether you're moving across the ocean like Magellan or outward bound from the grass strip to that fly-in over in the next county. The mariner works with two known quantities: velocity through the water and magnetic heading. For both, though, those two numerical quantities are not enough to ensure that your heading will produce the correct course. There's a missing third factor that all navigation systems solve. Mariners call it *drift;* pilots call it *wind.*

The basic geometric model for navigation is called the wind triangle in aviation circles, and every private pilot knows how to work one. You use it to translate heading and airspeed into course and groundspeed. Aviators may soar gloriously through the air, but navigation brings them back down to earth; it is the path over the ground that matters, after all, in navigation. The problem is that too many instrument pilots think they've left the wind triangle behind them when they move on to IFR navigation, that they've ascended into the world of moving needles, and that avionics obviate the wind triangle. The truth is that navigation aids are *not* the primary system by which an airplane finds the correct course under IFR; it's just that navaids are so conspicuous that they dazzle you with a delusion.

All IFR navigation is accomplished by dead reckoning.

There. It's been said. As a matter of fact, dead reckoning is the key link between attitude instrument flying and electronic navigation aids. The term "dead reckoning" describes a system for establishing position over the ground by use of a clock and a compass, and for some reason, it has acquired a bad name in aviation. Maybe people think it's too primitive, that it's for beginners but not for Real Pilots. The truth is that dead reckoning all by itself, without the help of electronics, is a remarkably accurate navigation method for airplanes, so much so that it forms the basic mathematical model for the computerized soul (you call it the algorithm, but that's one of those two-dollar words people hate) of the very fanciest nav systems in the larger airplanes, systems like inertial nav or Omega/VLF. All such nav systems revert to dead reckoning anytime they lose their ability to sense position from radio aids. Every one of those computers has been designed to come up with the missing third factor: the wind. Once a direction and value for the wind have been computed, the wind triangle is complete and airspeed/heading become groundspeed/course.

Two of the panel instruments that we've already talked about form the key link between attitude flying and navigation. The directional gyro (the compass, really) supplies magnetic heading; the airspeed indicator supplies airspeed. As an instrument pilot, your primary function is to aviate your airplane in such a way that its heading and airspeed provide you with the positional information you need in order to know where you are and to fly along the course you've planned.

When you learn dead reckoning as a VFR pilot, they teach you to compute a

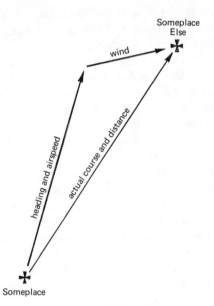

Attitude flying shows you only heading and airspeed, and in a situation with no wind, you'd be okay. But since the wind doth blow all the time, the trick to navigating in the real world is to pick a heading and to fly for a clock time that will put you at your destination and compensate for the effect of the wind. Wind is the unknown in navigation — forecasts are just estimates — and that's why you use navaids.

wind triangle from the forecast winds aloft, to apply magnetic corrections for variation and deviation, and to solve for a compass heading that should fly you directly to some checkpoint (or even the destination). Then you learn that wind forecasts are as truthful as matchbook advertisements — once you get up there and on your way, you find out your initial compass heading won't get the job done. When you fly VFR, you apply corrections through pilotage by comparing ground features with your sectional chart's depiction of your course. After a few minutes, you can determine whether you are left or right of your desired course (or, on those very rare occasions, dead on) and you can correct the compass heading to account for the actual effect of the wind.

All electronic navigation really does is to substitute certain transmitters at known locations for the visual references of VFR flying, and the entire history of aircraft navigation has been written in the evolution of improved transmitters.

Never forget that IFR navigation began before there were such things as electronic navaids for en route navigation. To substitute for all visual reference, pilots used only one very weak transmitter with a known location. It was called the magnetic north pole. They steered according to its directions, and aside from the nasty fact that steering an airplane IFR by a compass is a pain in the neck, magnetic north was not a bad navaid. Nobody ever shut it down, and with the exception of a few local anomalies, it had the widest coverage of any navaid ever dreamed up, including the radios that replaced it. To solve the navigation functions of direction and distance, the pilot needed only to use his magnetic compass for the former and a clock for the latter (airspeed multiplied by time yields distance).

The trouble with dead reckoning in an instrument environment is that you can't make corrections against any reference in order to solve for the real wind. Dead reckoning usually would bring you reasonably close to a destination, however, and you hoped for a hole in the clouds so that you might land. (There are still remote areas where IFR flying is accomplished in this time-honored way. Good thing *time* honors it, because *pilots* have come to expect more.) What if you could construct a second reference near the destination, a point in space that you could detect in some manner that wouldn't be affected by the weather? Why not use radio?

The first radio ranges provided not only a point in space, but four courses radiating from that point, usually 90 degrees apart. You could align one of these courses with a runway and guide an airplane to a landing. The ranges operated by providing audio signals through earphones, a pretty wretched system anytime a storm broadcast its own "top 40" while you were trying to hear an "A" or an "N" in Morse code through the crashing of static. So someone invented . . .

Direction finding, or radio compasses. A loop-shaped antenna could point fairly accurately at the course of a radio signal and thereby act as a kind of compass model. Build the transmitter anywhere you wanted one, and the receiver could point directly to it. Automatic direction finding — that's your ADF — relieves the pilot of the chore of turning the antenna manually, and that's the form in which we navigate using the signals from a nondirectional beacon these days.

NDBs have their disadvantages, though. Because of the low frequency ranges in which most civil models operate, they are vulnerable to atmospheric noise from lightning, and because electromagnetic radio waves in this part of the spectrum propagate differently over land from the way they do over water, ADFs are subject to a kind of refraction near coastal areas known as "shoreline effect." For these and other reasons, you'll find very few NDBs in use for en route navigation in the United States, although they continue to enjoy a prominent role in other countries. For the most part, your use of the ADF will be to navigate to compass locators, which are NDB transmitters located on instrument approaches. Also, at small airports without other approach aids, an NDB may offer the least expensive and simplest means to obtaining an approved instrument approach, and the number of such approaches listed by Jeppesen and the NOS is substantial. Unless you fly from such an airport frequently, though, you're likely to do very little ADF navigating. You probably will have to demonstrate a certain modest level of skill at determining bearing to stations and even tracking a course during your instrument flight test, and since

ADF navigating is fairly demanding of your talent for picturing the navigational situation, examiners find it a convenient way to test a student's abilities in general.

The great advantage to the ADF is that no matter where you are, if your receiver can detect the station, it will point out for you the station's bearing relative to the airplane; no other piece of avionic equipment performs such a function automatically. (Some VORs are now coming equipped with an automatic feature that allows you to push a button and obtain a bearing to the station without knob twiddling, but the fundamental principle of operation of the VOR is not bearing information but course-line depiction.) You will appreciate the ADF most when you are in the initial stages of an approach, with your VOR receivers tuned to approach navaids and the needles on those skewed to full scale while you are vectored by a controller into position to intercept a course line. Where the vaunted VOR boxes tell you nothing, the good old ADF will point doggedly at the compass locator at the outer marker and thereby provide you with the only information that helps you to stay oriented to the airport and approach.

Of course, you can navigate with an ADF. It's obvious that if you simply point the airplane in the direction of the beacon, using the receiver to ensure that the nose is always on center, you'll eventually fly directly over the top of the NDB transmitter. It's also obvious that this artificial north pole creates the same sort of geometry that the magnetic pole provides. Fly toward it, and there can be only one place you'll end up. Fly away from it, though, and unless you have a second reference, you could end up anywhere.

Therefore, to be of any real use, an ADF should be used in conjunction with a compass. Now all kinds of possibilities open up. With two known references — magnetic north and an NDB — you can find your position over the ground more precisely. If you can find a point, you can certainly describe a course, which is nothing more than several points connected together. With *two* NDBs and the right geometry, you can even triangulate.

Your ADF needle always *points* toward the station — that's the only function it can perform — and that's a nice feature if your only goal is to find the station itself; but when the time comes to describe a course line over the ground with any precision, ADF navigation demands some thinking and technique. The trouble arises when you plug the NDB system into a course line crossed by a stiff wind. Because the needle points only to the station and not to a course line, if you're blown off course by a wind, the display of the ADF alone gives you no ready way to find the course again.

Flying an ADF bearing to the station without respect to a second reference (the magnetic compass) in a crosswind will result in a course over the ground that resembles the one in the figure on page 00. Pretty sloppy, isn't it? You can see that the airplane's actual magnetic course varies wildly. Suppose there

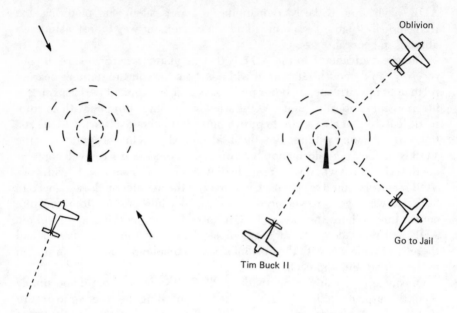

Fly to an ADF and you'll end up at only one place. Fly away from an ADF without a second reference (a compass) and you could end up anywhere.

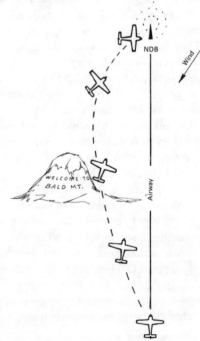

The trouble with "homing": Notice that even though the nose always points to the station, the course takes the long way around and just misses Bald Mountain.

were a mountain buried in the cloud somewhere along that long, curved line. Clearly, flying a steady ADF bearing with the needle centered while the compass wanders — a technique called *homing* — won't do for IFR navigation if you like assurances of terrain clearance and other comforting guarantees.

Obviously, for an ADF to describe a course line that's reasonably straight requires the adjunct of our second reference: the magnetic compass, or, as we use it, the DG. In order to fly a straight course, we have to learn an ADF navigation technique called *tracking*, which compensates for the effect of the wind (and it represents your first opportunity to fool around at the junction between attitude flying and navigation).

If an air traffic controller directed you to "track inbound on the 090 degree bearing to the Batten NDB," you'd be expected to establish a straight course inbound so that you describe a line along the ground that aligns with 090 degrees magnetic and ends at the beacon. Your antagonist is a wind from the northeast, but you don't know that yet. To get on that course line, you maneuver the airplane so that the DG reads 090 and the ADF needle points smack on the nose.

Take a look at your watch when you get started. Don't bother with the second hand; just note the minute hand for now. We'll say it's :05 past the hour. Time passes, as they used to say in the movies.

Now look at the ADF needle. If you've been doing your job and attitude flying according to the DG, you've been holding a course of 090. The ADF needle now points somewhere to the left of the nose, which tells you that the *station* is over there. Take another look at your watch. Let's say it's :08 past the hour. In about three minutes, you've drifted to the *right* of your course, and the ADF tells you that the bearing to the beacon is 20 degrees to the left of the nose.

Your problem: to get back on *course*.

An ADF indicator in its simplest form but with a rotatable card — you can adjust the circular plate holding the magnetic numbers so that your heading is under the index line — and with the indicator showing that the station is about 4 degrees left of the nose. Although this display allows you to rotate the compass card, most pilots leave the card just as it is here, and work in degrees left or right of the nose or tail while using the ADF to track.

The 090 degree course is also on the left side of the airplane but the ADF doesn't tell us exactly where. In order to get back on course, you'll have to turn the airplane to the left, but how much? You already know that if you merely point the airplane at the station, you'll end up homing, and that's no good. So a 20-degree left turn to align the ADF beacon with the nose won't do it, even though it looks so reasonable. (That's what stings a lot of pilots when they set out to learn ADF tracking. They teach us simple VOR nav in the private pilot course, and we get so welded to the idea of centering needles that we can't get out of the habit, even when we use a completely different system of navigation.)

We know we've drifted off course enough in three minutes to move the needle 20 degrees to the left. In order to get back on course, therefore, our left turn will have to be more than 20 degrees of heading change. If we were to steer to 070 degrees magnetic on the DG, we would align the ADF needle with the airplane's nose, which is insufficient.

Therefore, we *continue the turn past* 070 degrees on the DG to 050 — a turn of 40 degrees to the left, or *double* the original drift. Already, you can see a couple of very obvious things about ADF tracking:

1. To correct for drift, turn in the direction the needle points.
2. When you do, you'll be turning toward the wind and increasing the headwind component on the airplane.
3. With a higher headwind component, groundspeed will drop.

A. The situation at :05 tracking inbound on the 090 bearing to Batten NDB.
B. The situation at :08 shows the first effects of the wind as the airplane has drifted to the right of course, as indicated by the ADF's swing 20 degrees to the left of the nose.
C. How to make the wrong correction for an off-course drift: This pilot has turned left only 20 degrees and centered the ADF needle on a new magnetic heading of 070. This is nothing more than homing, and the airplane won't be back on the course centerline until it reaches Batten.
D. How to get back on the course correctly: Go past the 20-degree error and fly 050 degrees magnetic (that's a 40-degree correction) until the ADF needle shows an *equal* displacement in the opposite direction, in this case, to the right of the zero line. At this point all this pilot has to do is to make a 20-degree turn to the right and maintain a heading of 070 magnetic to take care of the wind correction. The airplane is on course now and should pass over Batten. ADFs, unlike VORs, don't show you course lines with their needles; they only point to the station. You have to assemble a geometric picture of the course line in your mind's eye using both the directional gyro and the ADF indicator.
E. Here's how the ADF and DG line up after the airplane is back on course. That's right; neither needle is centered on any course line. In fact, both instruments depict the airplane in its corrected situation. Whenever you navigate by ADF in a crosswind this will be the case. This pilot has the airplane pointed 20 degrees to the left of 090 to correct for wind drift. His ADF shows him that the station is 20 degrees to the right of the nose—which is exactly where it should be when the airplane is on the desired course, the 090 bearing.

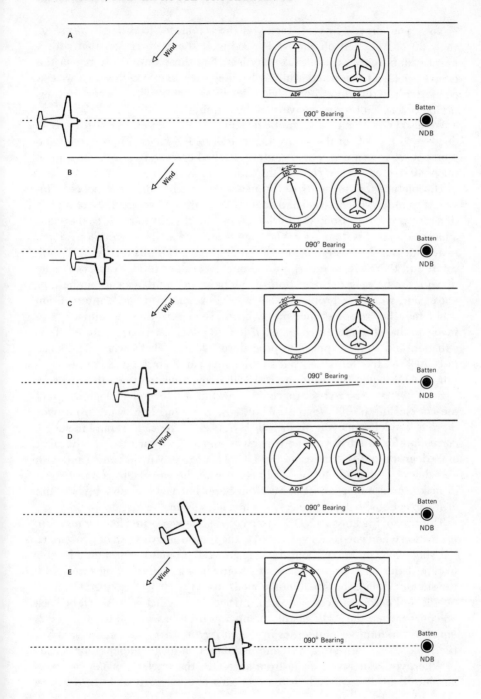

Now, you already know it took you three minutes to drift off course to show 20 degrees left on the ADF needle. It stands to reason that with a lower groundspeed, it can't take any less than three minutes to regain the course line! Isn't that wonderful? By keeping track of rate in this way, we can comfortably return to attitude flying, knowing that it will be a certain time period before another maneuver will be required.

We also know that on a magnetic heading of 050 degrees, our heading is 40 degrees to the left of the desired course of 090 degrees. Therefore, at the moment we regain the course centerline, the beacon ought to be on a bearing of 40 degrees to the *right* of the nose.

The picture this presents to the new pilot can be confusing, since it involves so much visualization and angle relationships. Because it's a relatively different procedure from VOR orientation and difficult to picture, there are almost as many "systems" for ADF nav technique as there are pilots. Some instructors teach a system using plus and minus quadrants on the ADF indicator, and there are formulas for adding ADF values to DG values to come up with the bearing to the station. If you have the mind of an accountant, go ahead and work the problems that way. For certain difficult interception problems, they're sometimes pretty good. Too often, these formulas get so involved that students and even experienced pilots lose their "picture" of the situation and simply apply the procedure by rote. That's when ADFs can bite. At least try this somewhat simpler method of tracking as an introduction to ADF navigation, and the rest will follow much more easily.

But I digress. You're back on course now but still turned 40 degrees left of the course line of 090 degrees. You know the wind is blowing from over there to the left somewhere, so your turn back to the right should include a *correction* for the drift you (now) know is there. Try a turn back to the right of 20 degrees to 070 degrees on the DG. That leaves you on course and with a 20-degree crab angle to the left to correct for the wind. Oh, you may have to make some minor adjustments from here, but you can now bracket the drift corrections, halving the correction angles as you've just done.

This way of tracking an NDB-based course line works just fine for en route navigation when you're not very close to the station and you're flying toward it.

Using what you know about tracking an ADF inbound, you should be able to repeat the same exercise in a situation wherein you're flying *away* from the station. Remember the fundamentals: turn the airplane toward the ADF needle and make your correcting turn twice the original drift. You'll be back on course when the ADF needle points toward a bearing off the tail that is *equal* to the number of degrees in the *correcting turn*. Now roll out to half the value of the correcting turn and maintain that drift compensation angle.

When you're heading *away* from an NDB, the tracking maneuver looks absolutely awful, especially when you turn to correct. It looks as if you're making the situation far worse than it was, and you must force yourself to picture the actual case. Think of where your airplane is with respect to the course line, and you'll do all right. Get yourself "VORed" and everything will go to pieces.

Occasionally, you'll be asked to intercept a bearing to an NDB. Big deal. Interceptions are nothing more than glorified course corrections, and you should treat them in the same way. If you have trouble getting oriented, it may help to turn the airplane to a heading that matches the bearing they want you to intercept. If it turns out you're not on the desired course line, reason tells you that you're at least *parallel* to it. After all, you're on the same heading. Just look at the ADF to determine whether the station is to the left or the right and proceed with the correction maneuver as if you'd drifted off course.

One thing about NDBs and air traffic controllers: the term *bearing* can describe a value that's either inbound or outbound with respect to the beacon itself, and that can confuse you unless you stay sharp. Usually, good controllers will define which they mean, inbound or outbound. Also, common sense tells you that most ATC directives are generally intended to speed you on your way. Any 180-degree confusion that sounds as if it might point you back toward the starting line is suspicious, and you'd ask the controller. (You would, wouldn't you?)

You'll sometimes hear pilots refer to the ground station as an ADF. You know better. The ADF is aboard your airplane. It receives the NDB, which is on the ground. You'll also notice that there are no rules calling for periodic checks of the ADF receiver equipment, which might mislead you into thinking that nothing ever goes wrong with ADFs. Goes wrong. Goes wrong. Goes . . . It's up to you to check your receiver periodically when you have a visual reference. The best you can do is to ensure that the needle generally points in the correct direction, using an instrument chart or a sectional to orient you to the transmitter position. There are, unfortunately, no handy methods for checking the *precise* alignment of the ADF.

Which brings us to our next navaid, the VOR.

The very high frequency omni range system, or omni, or VOR, is most adept at providing very precise *course* alignments. You will spend more time navigating on VOR signals than you will with any other single navaid. More than 1,000 VOR transmitters form a network of Victor airways and J routes that constitute the present system by which we get around under instrument flight rules. Area navigation, which ignores the structured airway system in favor of straight-line point-to-point travel, makes use of the VOR navaid system, nonetheless, and it's likely to be around for quite a while, at least until some form of positioning-satellite system takes over.

VOR provides some very significant geometric advantages over any of the

A typical lightplane omni display, complete with VOR, localizer, and glide slope guidance. The vertical needle is the course deviation indicator, or CDI, which you use for all lateral guidance modes like VOR nav and approach tracking. The OBS knob in the lower left-hand corner is the omni bearing selector, which turns to adjust the magnetic headings that appear around the periphery of the dial face. Right now, if the nav set were receiving (the OFF flags show it isn't getting either VOR/LOC or guide slope), you'd be on course (the needle is centered) at 236 degrees magnetic. The to-from indicator, which would appear in the lower center window where the OFF flag is now, would tell you whether 236 degrees was defined as toward or away from the station. The reciprocal of 236, which happens to be 056 degrees, can be seen handily at the circle on the bottom of the instrument face.

This model can be reversed for back course flying, and when it is, the BC light glows to warn you that needle sensing is backward. The warning light is there so that you don't inadvertently set it for back course when you want to fly a front course, and vice versa.

preceding navaid systems, and furthermore, its mode of presentation makes life very easy for you, the pilot. Instead of the NDB's point-in-space "pole" that radiates a homing signal, the VOR broadcasts 360 course lines that radiate from it like the spokes of a wire wheel. Instead of presenting you with bearings to a station, your VOR display shows you 360 *radials* that correspond to the 360 degrees of the mag compass. The receiver display also identifies exactly which radial your airplane is positioned upon at the moment, without any need to consult a magnetic compass for help. Thus, the VOR can give you a course to fly by centering a needle on its receiver display or it can give you position information from one side of your course with very little fuss on your part. Whereas the NDB's signal carries no identifying positional information, the VOR's signal delivers a phase difference to the receiver that translates into a magnetic course.

The ADF presents its information to the pilot with respect to the direction in which the airplane is pointed; its arrow aims at a single point, after all — the NDB station. Since the VOR display presents the pilot with references to radials, or lines in space, it ignores airplane heading entirely. The VOR's *course deviation indicator* (CDI) — a needle that deflects to the left or right of a center index — tells you only where you are with respect to the radial you select in the *omni bearing selector* (OBS).

If you were to park an airplane on an airport and dial in a VOR station, then twiddle the OBS until the needle centered, all you would really know was that the airplane was positioned over that particular radial. You can rotate the airplane in any direction or orientation you want to, and nothing will change. The VOR receiver on the panel is telling you all that it can: "This airplane is currently sitting smack on top of the 00 radial from Buckminster VOR." The reason is that Buckminster's signal is merely information that defines a course line in space.

This course line, or radial, can have two definitions. Let's say the radial you selected happened to be the 090, or direct eastbound course, *from* Buckminster. You could just as easily define it as the 270 radial *to* Buckminster, and in fact, if you were to twiddle the OBS again until 270 was under the index, the needle would center again. As you can see, each radial has this ambivalence to it, and the numbers 090 and 270 in this case, are known as *reciprocals*.

To eliminate the confusion of radial ambivalence (sounds like deviant behavior, doesn't it?), the VOR receiver will display either radial for you and define it as "to" or "from" the station. The device that provides this service is called the *sense indicator*, usually dubbed the "to-from indicator" by most pilots. This can be either a pair of arrows, one pointed up and one down and meaning to and from (but not both at the same time), or it may be in the form of a little window with the word "TO" or "FROM" appearing as appropriate. The sense indicator will *not* determine for you that the airplane is *flying* to the station; it is merely part of the definition of the radial you tune. Another way to demonstrate this most mysterious characteristic of the VOR for good and always is to take an airplane with a VOR receiver and fly it a good 25 NM from a transmitter, then center the needle on any radial and fly a tight 360-degree circle. Notice anything funny? Nothing changes on the VOR display! Just as it did with the parked airplane, it told you that the airplane was positioned in the middle of a defined course.

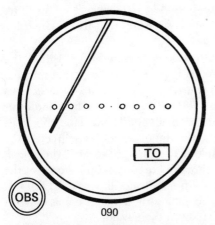

090

How to operate a VOR receiver: After tuning the station and identifying it by its Morse code transmissions, adjust the omni bearing selector (OBS) to the magnetic radial you want. Here, we've selected 090. The course deviation indicator (CDI) is a needle showing us that the radial we've tuned is close by. But depending on which way the airplane is pointed, the course may be to the left or right of the airplane.

The little window that tells us the 090 we dialed is TO the station merely eliminates half of the radial that leads FROM the station at 090 degrees. It does *not* tell us that we are flying to the station; only the compass can tell us that.

There is no connection between your compass and your VOR receiver, so VOR navigation, too, requires the adjunct of a compass. Here's the rule: *If the radial you select matches the compass heading you're flying* (within a few degrees), *the sense indicator is telling you the whole truth.* An example to help clarify the rule: You're flying outbound from a VOR and you center the OBS needle and discover you're on the 090 radial; your compass agrees with that — 090 degrees (there's no wind). Your sense indicator will read FROM and it will be correct; you *are* flying away from the station. Now turn the OBS until you have the reciprocal course selected with the needle centered. Don't turn the airplane, however. The radial should now be 270 and the sense indicator will read TO. It's the same radial, different definition is all, and the airplane is most certainly *not* flying at 270 degrees toward the station. The point is that it's up to you to fly the airplane in the right direction, and further, you have to provide wind corrections to keep the airplane on the course you select.

For some reason, that responsibility gives pilots fits too. A friend of mine once watched an experienced pilot fly to a VOR by taking a heading and centering the needle — as he was supposed to do. But as this guy flew along and the needle moved to one side or another, indicating that the course was over here or over there slightly, he'd readjust the OBS to recenter the needle! By so doing, he managed to *home* to a VOR — exactly what the system was designed to avoid.

The beauty of the VOR course display is that as you drift slightly off *course*, the CDI needle (assuming you've defined a radial that matches your heading) literally points you in the direction you should fly to correct and displays your angular *course-line error.* Turn the airplane toward the needle until the needle recenters and you're once more back on course-line center. Even if the airplane has to crab 20 degrees for a wind correction, so long as it is positioned on the center of the course line, the VOR display's CDI will remain centered. VOR is far easier to use than an ADF, and that's why it has largely taken over for en route navigation.

When you fly IFR, you use the VOR in very specific ways. Either the ATC/airway system will provide you with a specific course to dial into the OBS, after which you simply track that course, or it will clear you *direct* to a particular navaid, in which case you turn the OBS until the to-from indicator says TO and center the needle. Read the magnetic heading on the VOR display and that's your magnetic course *to* the station. If you pass directly over the station, you'll start flying outbound along the opposite spoke of the wheel. The needle may wiggle a bit as you pass through the cone of confusion over the top of the station, then stabilize as the to-from indicator flips over to FROM. That moment is called "station passage," and it may be a time check as well as a cue to reset a new course outbound away from the navaid.

To track a course, determine first whether the station you want to use is in front of you or behind you. If it's in front, turn the OBS until you get a TO indication, then dial in the course and maneuver the airplane in the direction the CDI needle points until it centers. You're on course to the station. Set up a magnetic heading that coincides with the number you've tuned on the OBS and turn the *airplane* to make course corrections and track to the station. Leave the OBS knob alone; it's set for your desired course, and if you use it to center the needle, you'll be on some other course line. To track away from the station, the only thing that changes is the to-from arrow; it'll read FROM.

To orient yourself with respect to the station requires that you be aware of your airplane's heading. If your compass and the number you dialed with the OBS knob are the same, the to-from indicator will tell you whether the station is in front of you or back there somewhere.

Occasionally, ATC directs a pilot to "fly inbound on the _____ radial." This too sometimes creates confusion for some pilots.

Unlike the word "bearing" as it's used around NDBs, the word "radial" means only an *outbound* value in ATC talk. Thus, the 090 radial can *only* be the line extending due east from a VOR. If your airplane were flying westbound in airspace to the east of a VOR and you were told to fly inbound on the 090 radial, you would tune the reciprocal of that radial, or 270 degrees. Most VOR displays either show you the value for the reciprocal in tiny numbers beneath the omni bearings themselves or they'll have a separate little window on the opposite side of the dial. (Some of the newer ones even allow you to push a button and automatically set the reciprocal into the selector. It saves twisting the OBS knob all the way around to the other side.)

Occasionally, you use VOR radials to the side of your course to identify reporting points or navigation fixes. There's a very simple way to orient yourself using the VOR radials off your course. Always tune a "from" with the OBS and then the value of the radial you'll be using. Usually, the needle will swing to one side or the other unless you're flying on that radial at this particular moment. Is the reference VOR on the left or right of course? If the needle points in the direction corresponding to the side of your course on which the VOR is located, *you have not passed the radial yet.* Draw some situations out on paper to see why this is so; at moments when you have trouble constructing your picture of the situation, rules of thumb like that one can be handy. Visualizing where you are is much better, though, and after a while, you'll have enough practice with the omni receiver and display in your airplane to draw a quick mental picture anytime you tune the VOR.

Distance-measuring equipment (DME) presents you with very accurate information regarding your slant-range distance from a DME facility. The equipment works by measuring the time lag between a transmitted pulse from your airplane and a return pulse from the ground station. Because DME ena-

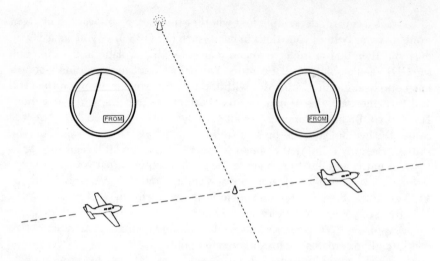

To use a VOR that defines an intersection, tune the correct radial FROM the station. Determine which side of course the station is on (left or right), and if you have not yet reached the intersection, the needle will point in the corresponding direction.

bles you to know your position very accurately, it has a separate equipment code in flight plans, one of three pieces of equipment (the other two are the transponder and area nav computer) that get their own codes. If the DME should malfunction anytime you're operating IFR when you've filed the DME equipment code, you must report the DME inoperative to ATC so they can make adjustments. It may be that they had you plugged into a DME arrival procedure or may have assigned you a DME hold and they'll need to know if you don't have one working.

Pilots without DME estimate distances from fixes by timing, which can be inaccurate without reliable wind information. Unless you understand *slant range*, however, you may be just as inaccurate in your estimate of your position *with* a DME. Slant range simply means that the DME derives the distance on a direct line in three dimensions, through the air, between the airplane and the ground station. That line always describes a *slant* unless for some strange reason your airplane is at exactly the same altitude as the elevation of the DME station. You use the DME information as if the difference were minimal between your slant range and the *real* distance you use for navigation computation — the linear distance over the ground from a point directly beneath your airplane to the DME station. The slant range and your ground distance form two sides of a triangle, while your height above the ground datum forms the triangle's base. If you are flying at 10,000 feet, you are almost two miles above the ground datum line. (See figure, page 00.)

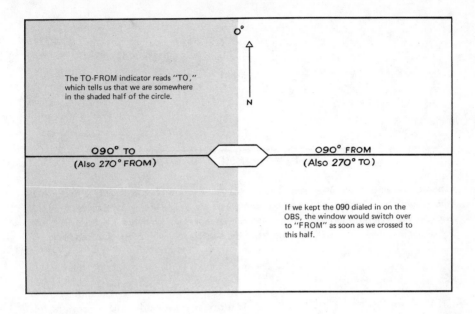

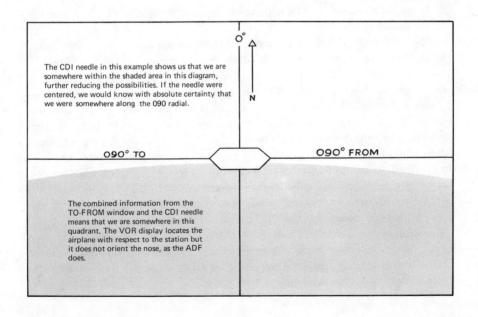

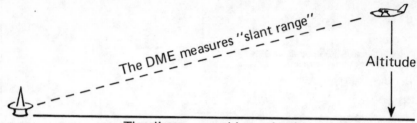

The distance used in navigation

The trouble with DMEs: The "slant range" they measure is always slightly greater than the ground distance you use for navigation purposes. When you're far enough from the station, the difference is insignificant. The rule of thumb says a mile for every 1,000 feet of altitude will be sufficient distance from the station to give acceptable readings.

Obviously, as you fly over the station, the DME will tell you that you are still two miles distant. It's not lying. You *are* two miles distant — straight up! Trouble is, our math model for navigation exists on chart paper, in a geometric plane, and the height/slant range "error" introduced by the DME must be considered. As you can see, the distance error varies in its effect. If you're a hundred miles from the station, your 10,000-foot error is a very small percentage of the distance being measured.

Therefore, remember: DME slant-range error is significant when you are closest to the station.

Most DMEs display these important data: distance (in nautical miles) from the station, time (in minutes) to the station, and/or groundspeed in knots. In many cases, time to station (TTS) and groundspeed are simply two different scales that can be interchanged underneath a needle or otherwise moved internally. Both of the latter values are obtained instantly by the unit by electronically measuring how fast the distance to the station is changing with respect to time (there's a kind of clock in the DME). Usually, ATC controllers know your groundspeed accurately enough for their purposes, and so the timing is likely to be of more interest to them. It's those situations wherein you are quite suddenly asked to estimate your arrival at a fix that cause you to love a DME, and most instrument pilots would probably admit (maybe off the record) that estimates are half the reason for owning one. Unless you are navigating on a course that leads directly toward or away from the DME station to which you have tuned, remember that the changes in distance that it sees are angular, not straight-line, and that's why you can't make use of off-course time-to-station data.

Chances are that if you operate a DME over a period of time, you'll tune a navaid one day and wonder why the DME won't "lock on" to the station. The cause may be some malfunction in the station, but before checking anything else, refer back to your navigation chart to be absolutely certain the ground station you've tuned is equipped with DME. The term "VORTAC" has come to mean a DME-equipped VOR navaid even though it isn't strictly accurate. The "-TAC" in "VORTAC" actually refers to TACAN, a military version of VOR/DME that provides azimuth and distance information to the troops in a single receiver. Charts make no distinction between DME and TACAN in their symbology, however, and if you hear the word "VORTAC" in the verbal identifier, your DME ought to work. The overwhelming majority of the VOR navaids in this country are actually VORTACs, and if you're human, sooner or later you'll tune the exception — the VOR-only station — and wonder why your DME won't play. Check the chart. Bet you a five it's a non-DME navaid. If the chart says it's a VORTAC, then listen for the DME identifier, which comes up less frequently than the VOR code and at a higher pitch tone. Like the VOR system or any very high frequency radio facility, DME (which operates in the even higher UHF band) operates on the principle of line of sight. For all practical purposes, if you can't "see" the VOR, the DME won't play either.

DME is not a required piece of equipment for IFR flight. It is required at high altitude, but that's an airspace/ATC requirement. DME may be required for certain approaches, but for lightplane en route navigation, it's really a convenience. When ATC asks you for a position report — during, say, a handoff between two radar controlling facilities — you can make identification of your airplane easier by telling the controller, "_____ miles *DME* southeast of Albuquerque VORTAC." Add the letters DME after your mileage and that tells ATC to look *exactly* where you said to look; their radarscope displays a very accurate mileage-distance map.

Quite frankly, the en route portion of the flight is the most leisurely segment of any IFR flight. It begins at the transition from the departure procedure to an airway and ends as soon as you enter the arrival segment prior to landing. En route segments are characterized by straight-and-level flight. For most pilots, it's a time to switch on the autopilot, make fine adjustments to the powerplant for maximum fuel efficiency at the cruise altitude, and *plan ahead*. Toward the end of the en route segment, prior to the added workload of descent maneuvers, you should begin to prepare for the approach by getting your charts ready. You already have current weather reports for the destination airport (ATIS or automatic terminal information service), and those tell you the surface winds. (Bet you can guess from that information which runway is likely to be in use. I can count on one hand the times I've been asked to land downwind or in any major crosswind.)

During the en route segment, you're likely to be at the highest altitude you'll reach during any IFR flight. It's a good time to call Flight Service stations for weather updates and for using oxygen if you'll need it. During en route segments, talk between you and ATC is at its minimum. Most people hate moving an oxygen mask around to talk through a microphone, so why wait to use the tank just before the approach on the theory the O_2 will perk you up and have you fresh for the approach? That's also the time when the chatter picks up, and oxygen doesn't work instantly anyway. It sort of soaks into your body over a period of a couple of minutes. Use it prior to the descent and you'll have the mask out of the way to talk while you're headed down toward denser air anyway.

En route segments are a time when the attitude flying chore is lightest, and the overlying task of navigation really amounts to dialing in various courses on the OBS knob and then tracking the course. ATC expects you to fly within the confines of an airway if you are navigating on one, and since airways are only 8 NM wide, that requires some attention to the CDI needle, particularly when you are a long distance from the station. Many pilots are surprised to get a call from center informing them that they are more than four miles off the airway centerline. Since only the protected airspace within the airway boundary guarantees terrain separation, ATC has good reason to worry about you. Remember too that you are expected to reset your altimeter to the nearest reporting station's barometer within 100 NM. Effectively, it means that you have to reset the altimeter *every* 100 NM; after a while, you'll get into a rhythm and realize that in airplanes, distances always equate to time. Most lightplanes cover 100 NM in an hour or less. The rule says your station setting must be "current," but it doesn't tell you what that means. A good source for altimeter settings is the automatic terminal information service (ATIS) tape at ATC airports. Too many pilots think that because ATC folks are in the habit of furnishing altimeter settings during handoffs and sometimes as disguised com checks, that pilot responsibility has been relieved; not true.

Your tracking responsibilities are even more important, and aside from keeping the course needle centered (the airway system is designed to accommodate 2.5 degrees of pilot deviation, but that doesn't mean you ought to use it all up), the only area for potential error is setting the incorrect course into the VOR receiver or simply steering to an incorrect airway. At VORTACs where several airways intersect, it pays to be a little more careful to ensure you have the correct magnetic course for the airway you've been cleared to. For one thing, it's embarrassing to have a controller call you and remind you that you're supposed to be on Victor 107E instead of Victor 107. Being embarrassed is nothing, however, compared to what could happen if ATC fails to notice.

A flight director in the en route environment tunes itself to "low gain," which is a way of saying that it makes its corrections very gradually. A smooth en route maneuver makes for comfortable passengers, but there's at least one situation wherein you'll want to avoid the extreme: turns in airway doglegs or over navigation fixes when *overshoot* might carry you out beyond the protected airway boundary. Granted, this is something that is more of a worry to pilots of airplanes with higher performance, but bear it in mind. The airway does not get any wider around a turn; above 18,000 feet, ATC does provide additional IFR separation protection for turns made at higher airspeeds, though.

Now that you're aware of that little warning, don't overreact on airway turns just because the VOR needle slams over to one side or the other and appears to be telling you that the course centerline is miles away. Station passages that involve course changes are a good time to take rate of change into consideration and adapt your navigation to suit the situation. In short, it's a good time to revert to dead reckoning. For one thing, your VOR is no longer delivering reliable information, and if it were, the geometry close to a station can mislead you if you fly only the VOR needle.

By the time you've been en route awhile, you will have noticed that the VOR course you've selected may only approximate the magnetic heading you have to fly to track the course. That's one nice thing about VOR navigation: built-in crab angle detection. If it takes you a magnetic *heading* of 035, say, to maintain a VOR magnetic *course* of 043 degrees, you are quite correct in concluding that you're flying through wind that's on your left wing and with enough of a crosswind component to force you to fly an 8-degree wind correction angle. (The wind is actually seldom steady enough to allow us to talk that precisely, and you'll actually be attitude flying with occasional reference to the CDI needle and making constant corrections as the wind changes. So what we're saying in this example is that over the long term, you notice that an average of about 8 degrees seems to be your correction angle.)

As you approach a course change, there is no reason to suspect that your drift correction will differ radically from what it's been for the last two or three minutes. Thus, to turn to a new heading when crossing a navaid, simply ignore the VOR for a while and steer to a new heading that includes a correction for drift added to the desired magnetic course. After two or three minutes, the VOR needle should begin to come back toward the center (assuming you remembered to select the new course on the OBS), the navigation signal will become more reliable, and you can again return to VOR navigation as a supplement to dead reckoning.

The truth is that you've been dead reckoning, at least in terms of steering, all along. If you are navigating en route according to a VOR in the correct way, you should be steering a *DG heading* that keeps the CDI needle centered. It can't work any other way. You can't steer by the CDI alone — try it sometime — because it presents varying needle deflections depending on your distance

from the station. Magnetic compass headings may change a little due to variation, but for steering purposes, their *rate of change remains steady, whereas
the VOR's response varies.* Understanding that one point is the key to understanding navigation, and when we get into approaches, you'll see the extreme
case — on the ILS — for using the DG as the primary steering reference.

(When I first started reading IFR navigation charts of the en route variety, I
couldn't figure out why an airway leg that began at one VORTAC at a particular magnetic course would terminate at the next VORTAC up the line on a
slightly different magnetic course. If an airway is a straight line, how could the
magnetic course be different at two ends of the same leg? The answer is simply
that VORTACs are oriented to magnetic north, and since magnetic north dif-
fers from true north by the local *magnetic variation*, two stations a hundred
miles apart or so may very easily lie in areas with two different magnetic variations. At the navaid changeover points marked on the chart, you not only
change frequency, you may also change magnetic course to account for the difference in magnetic variation.)

Simple autopilots that steer according to a CDI needle's left-right commands mislead us into thinking that all autopilots really fly the needle and not
the compass. Navigation computers and more advanced systems really weight
their steering commands very heavily toward heading, however; the simple
autopilots substitute a similar form of damping to smooth out the steering
commands — if they didn't, you'd be mighty uncomfortable following every
course line wiggle. Pilots who fly a lot in mountainous areas know that terrain
reflections can distort VOR course information and produce "scalloping," or a
waviness in the radial. When that happens, the thing to do is to "weight" your
steering·for heading; watch the CDI needle for an average of the course
centerline, but hang onto magnetic heading like a hungry dog. You'll soon
pass through the area where reflection is disturbing the needle and once again
regain a steady nav signal.

In practice, you'll be integrating two sides of your navigational house in one
instrument when you steer. Directional corrections for attitude flying take
you to the DG during the en route phase. Navigational corrections are also re-
solved at the DG in the form of small heading changes. By the simple act of
flying the airplane so as to track a VOR course and keep the wings level, you'll
complete the third vector in the wind triangle — the wind — without even
realizing it and with very little effort.

Unless you have to hold.

Everyone hates holding, and there's very little to be said in favor of it, since
it delays you, interrupts the flow of your routine, and raises anxiety. It also
adds work, and in today's ATC environment, unless you fly long distances and
often, into some of the more congested areas, the word that you're going to
have to hold can come as quite a jolt.

When ATC issues you a clearance with a limit short of your destination airport, they're supposed to issue you an additional clearance prior to the time you arrive at the clearance limit. But what if that additional clearance never comes and there you are at the limit? Within three minutes of arrival at the clearance limit you are expected to reduce speed and prepare to enter a standard holding pattern using the limit fix for a holding fix and the en route course line for a holding course. In other words, unless there's a nonstandard hold depicted at the clearance limit, just fly right into the turn of a standard holding pattern. You do this *on your own* — while you climb all over the radio trying to get a further clearance. It's one of the easiest ways you'll ever enter a holding pattern, since you're really all set up for it. Just cruise until the needles say you're at the limit fix, look at your watch, and start the turn. You should report the time you entered the hold just as you do if you're assigned a holding pattern. Even if you're on radar, you'll get their attention that way.

All other holding entries will be made in accordance with ATC clearances, and will include all the necessary information you need to establish the pattern. If an essential piece of information is missing, however, you have to ask for it, and what's usually missing is an *expect further clearance* (EFC) time. You may wonder why EFC times are so important, and the answer is this: what would you do if you lost communications while you were holding? The EFC time tells you when, in the event of lost com, to leave holding and proceed to the next clearance limit or your destination.

The only difficult part about holding is figuring out how to enter the pattern according to the FAA's "recommended" entry turns when you're coming at the fix from upside down and backward.

First, understand that the recommended entry turn is meant to apply within 5 degrees of the headings depicted, so they're telling you it's okay to fudge a little. Second, the reason for going to such lengths to depict proper entry turns is to keep you inside the protected airspace within the holding pattern. If you have slowed to a comfortable holding speed in a lightplane, it is extremely doubtful that you will barrel through there so fast that you'll nose your way into dangerous airspace; there is a buffer around holding patterns to accommodate high-performance airplanes that fly much longer legs in one minute than you will.

Some flight test examiners like to insist on the depicted entry turns, and my guess is that they use it as another test of your ability to visualize complex navigation situations and keep your cool. For them, use the DG method if you can get it together in your head. On those occasions when it all happens so fast you know you haven't a prayer of figuring out the proper entry in time and you merely want to enter the hold with the minimum fuss and do nothing dangerous, use the stone-ax simple method. Either way, first you have to make sure you'll be entering the proper pattern.

An ATC holding clearance is supposed to tell you the following details:

— one of eight compass points describing the direction in which your holding course lies from the fix

— the name of the fix (it helps, by the way, to scan en route charts when you have nothing else to do and just read quickly the names of intersections around you; over the radio, some of the new five-letter names don't sound the way they're spelled)

— the holding course definition (could be an airway, a radial, or just a heading)

— leg length in miles (only for DMEs or nonstandard timing)

— "left turns" if they want a nonstandard pattern (they don't say anything if they want a standard pattern)

— "expect further clearance at _____ after the hour" (and don't let them forget it; they help *you* to remember things)

En route holding is usually performed at the altitude you've been assigned, since that's already reserved for you, so terrain clearance is generally implied in your altitude of MEA or higher. For airspace protection, however, maximum speed limits are given to confine the holding pattern itself to reasonable dimensions. So try to keep your Skyhawk down below 175 knots; I know it's an effort. Think of the poor SR-71 pilot, who has to drag it along at a mere *310 knots* indicated. You are expected to holler to them if wing ice or some other condition requires you to fly faster than the maximums.

Here's how to diagram a holding pattern when you are assigned one: Use the chart. Write directly onto it with pencil. First, find the fix. Now draw a line from the fix outward along the compass direction they gave you. That should correspond to the *inbound* holding course, or check to see that it does. The inbound course always flies directly over the fix. Now place your pencil at the *outer* end of the line and "*fly inbound*," making a right or left turn, for standard or nonstandard patterns, and shazam! Just fill in the other turn. It will do no good to fly a perfect entry turn into the wrong holding pattern.

A friend of mine, an experienced pilot, was taking a check ride in the left seat of a three-holer Boeing simulator once and a very senior line captain was right-seat crew. The FAA man giving the test called for a holding pattern over a nearby intersection, and the pilot blanked out. He not only couldn't envision the entry — he couldn't even get a clear picture of the holding pattern! In a moment of brilliance, he turned to the copilot and simply said, "Figure me the correct entry into the hold and advise. I'm reducing speed now." Turned out that was the right answer. Not only was the entry and hold perfect, but he got an A for crew coordination.

You won't be that fortunate; we're not allowed to take tests with senior line captains for copilots, so you have to do it all yourself. Take comfort in the fact that holding is the only interruption to the smooth continuum of en route flying, and it doesn't happen that often. There may come a day when you'll welcome a break in the routine. And in a way, if all of us were called upon to fly holding patterns more often, they wouldn't seem like such a crisis when they pop up.

IFR Notebook

IFR RADIO NAVIGATION AIDS

1. **Very High Frequency Omni Range (VOR)**
   Frequency: 108.0 to 117.95 MHz
   Line-of-sight ranges by class of facility:

T	*terminal* class, about 50 watts, 12,000 feet and below	25 miles
L	*low-altitude* en route class, 200 watts, below 18,000 feet	40 miles
H	*high-altitude* en route class, 200 watts, below 18,000 feet	40 miles
	14,500 feet to 17,999 feet (in the continental U.S.)	100 miles
	FL180 to FL450	130 miles
	above FL450	100 miles

To obtain maximum usable distances between VORs (for filing *direct* between two navaids), add the above factors for each facility for the total usable distance.

Airway dimensions: 4 NM each side of centerline, except beyond 51 NM, at which point the airway is described by a sector defined by radials 4.5 degrees on each side of centerline.
   Lower limit is 1,200 feet AGL or as stated.
   Upper limit is 17,999 feet AGL or as stated.

CDI needle displacement value: One dot equals about 2 degrees or approximately 200 feet for each NM from transmitter.
   (Example: At 20 NM from the station, a one-dot displacement would equal 4,000 feet — about ¾ mile — from centerline.)

VOR accuracy:
   Theoretical accuracy — ±1 degree
   En route radials — within 2.5 degrees of published
   Approach radials — within 1.5 degrees of published

Identification:
   Three-letter Morse code every five seconds
   or, code and voice, with voice every 15 seconds

Communication:
   Airplanes transmit on 122.0 MHz, receive on the navigation receiver through the ground navaid frequency; turn volume up, ident off.
   When using a VOR for voice communication with Flight Service, there is no interference with its navigation function.

VOR time-to-station formula:

$$\frac{60 \times \text{minutes flown between bearing references}}{\text{number of degrees between bearing references}}$$

VOR distance-to-station formula:

$$\frac{\text{true airspeed} \times \text{minutes flown}}{\text{number of degrees between bearing references}}$$

Quick rule of thumb for time to station:
   One tenth of the time in seconds to fly through a 10-degree bearing change equals the time to the station in minutes.

VOR vertical angle coverage limitations (refer to AC 00-31):
   Azimuth signal information is satisfactory up to an angle between the radio horizon and *not less than 60 degrees* elevation angle.*

Suppression of ident signals:
   When voice communication is being transmitted on VOR facilities, the code identification is *not* suppressed; however, voice identification signals *are* suppressed.

*In practical terms, at altitudes less than 20,000 feet, this cone should never exceed an effective radius of two miles. You fly two miles in about a minute, in most lightplanes.

2.   **Non-directional Radio Beacons (NDBs)**
      200 to 415 kHz, one-kHz separation
      carrier modulated with 400 Hz or 1,020 Hz audio tone for ident. The NDB is NOT limited by line-of-sight rules.

Usable distance, by class:

Compass locator (less than 25 watts)	15 miles
MH (less than 50 watts)	25 miles
H (between 50 and 1,999 watts)	50 miles
HH (2,000 watts or greater)	75 miles

Usable distances are the same for *all altitudes*. Some H class facilities may have usable distances of less than 50 miles.

Identification:

Three-letter Morse codes, except for compass locators.

Voice transmission:

Radio beacons are capable of voice transmission (and may also transmit scheduled weather broadcasts) unless their class designator includes the letter W, which means "without voice."

3. **Distance Measuring Equipment (DME)**

Frequencies are generally paired with VOR, localizer, and ILS (instrument landing system) operational frequencies and listed by channel numbers in accordance with the reply frequency of the ground facility. Channels 1X and 1Y through 16X and 16Y are not paired with any frequencies in use at this time. The lowest frequency in use, 108.5 MHz, corresponds with channel 17Y, which interrogates on 1,041 MHz and replies on 1,104 MHz. Each numbered channel is paired according to X and Y reply frequencies, which are separated by 126 MHz. At a VOR frequency of 117.95 MHz, the highest in use at this time, the interrogation frequency is 1,150 MHz and the reply 1,087 MHz. Channels 60 through 69 are not assigned corresponding VOR frequencies.

Accuracy: Slant-range distance is provided up to 199 NM at an accuracy of better than .5 miles or 3 percent of the distance, whichever is greater. The crossover point at which 3 percent becomes greater is 16.7 miles (and you wonder why they don't just say so). Slant-range error is negligible at more than one mile distance for each 1,000 feet of altitude.

Identification: DME components colocated with VOR and approach facilities transmit coded identifiers once for every three or four VOR or approach-aid ident code transmissions. If the nav facility is inoperative, the DME alone will transmit once every 30 seconds on a frequency of 1,350 Hz audio, a tone which is higher in pitch than a VOR's 1,020 Hz tone.

DME geometry:

Distance to station: Always valid but *slant* distance, not ground point-to-point distance

Time to station/groundspeed: Valid *only* when aircraft is flying directly on course to or from a DME station, not when DME is located to either side of course line.

4. **Omega/VLF** (Note: These advanced navigation systems are included here for those pilots who wish to refer to the data.)

Omega stations: Norway, Liberia, Hawaii, North Dakota (U.S.A.), La Réunion, Argentina, Trinidad, Australia, and Japan.

VLF stations:	Annapolis, Maryland	NSS	21.4kHz*
(with idents and	Cutler, Maine	NAA	17.8*
frequencies)	Balboa, Canal Zone	NBA	24.0
	Jim Creek, Washington	NLK	18.6*
	Lualualei, Hawaii	NPM	23.4*
	Northwest Cape, Australia	NWC	22.3*
	Rugby, England	GBR	16.0*
	Yosami, Japan	NDT	17.4*
	Helgeland, Norway	JXN	16.4
	Anthorne, England	GQD	19.0*

Omega frequencies: 10.2 kHz, 11.05 kHz, 11.33 kHz, 13.6 kHz in a rotating cycle of timed transmission with repeats every 10 seconds, at about 10 kilowatts.

*Used for navigation.

In addition, "nav frequencies" for certain stations are broadcast.

U.S. Coast Guard taped status reports on 202/245-0298; also broadcasts on WWV, Boulder, Colorado, short wave transmission at 16 minutes past the hour or WWVH, Hawaii, at 47 minutes past the hour. Notams concerning operation of Omega/VLF transmitters available through Flight Service.

Accuracy: About ±2 NM

Navigation on the Navy VLF communication system alone is not approved for IFR. VLF is considered a supplement to Omega.

5. *Other systems:*

Inertial guidance systems

Doppler

Loran C

These and certain other long-range navigation systems are normally installed

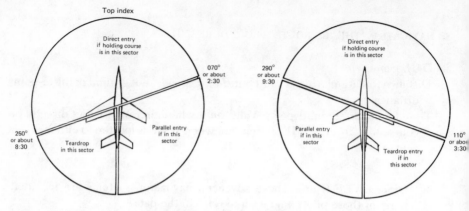

How to use your DG as a visual aid for holding pattern entry turns:
For Standard Turns: Find the holding course heading on the DG. Which one of the sectors shown here does it lie in? Fly the entry turn that corresponds to the sector.
For Nonstandard Turns: *Stone-ax-simple method* for saving your life when you can't think of the right entry to make: Cross the fix and head outbound for one minute on a magnetic heading that's the reciprocal of the holding course. Check the CDI needle. Turn toward the course.

only in transport category aircraft or other large aircraft and operate on principles that are distinct from the ground-based navaid system used for conventional IFR navigation.

About these navaids:

Terminal class navaids are intended as approach aids into specific airports and therefore have severely limited range. On charts, they are symbolized by a somewhat smaller VOR compass rose, whereas the low- and high-altitude class VOR is represented by a larger symbol. High-altitude navigation en route charts also distinguish between low- and high-altitude VOR stations, but low-altitude charts do not.

Note that use of ADF equipment to receive commercial broadcast signals for navigation purposes is not approved for IFR. Commercial broadcast transmitters are not supervised by the FAA and may have limited licenses that require them to cease operation at certain hours. Only stations that have demonstrated "circular radiated power" (a characteristic of the antenna array) are useful with an ADF for VFR flight; the Jeppesen J-Aid lists such stations.

For all practical purposes, you need not concern yourself over frequency pairing of DME equipment with VOR channels. It's sufficient to understand that when you tune your DME to what appears to be a VOR frequency on the selector, you're actually tuning in a much higher frequency. Some DMEs "slave" to the VOR and automatically tune the correct paired channel for the frequency tuned in the VOR receiver to which they are linked.

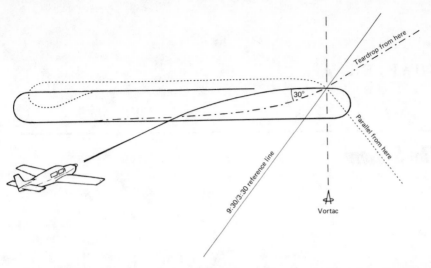

A view of a holding pattern from slightly above and in an airplane about to make a direct entry. Generally, if you're approaching from the holding pattern side of the fix, a direct entry is correct. From the other side, it's either a teardrop or a parallel, whichever takes the less amount of heading change.

Holding Pattern

QUICK REFERENCE

Timing: At or below 14,000 feet   one-minute legs
Above 14,000 feet   1½-minute legs

Reduce speed to holding maximums *within* three minutes of entering hold.

Report: TIME
FIX   *only* when you're *not* in radar contact
ALTITUDE

Clock: *Time* initial crossing of fix
— over or abeam fix outbound, whichever occurs later; outbound, when turn completed, if abeam cannot be determined
— 3 degrees per second turn recommended
VOR: "to-from" reversal
NDB: needle reversal

Maximum holding airspeeds:
Propeller aircraft   175 knots *indicated*
Turbojets (all *indicated* airspeeds)
minimum holding altitude to 6,000 feet   200 knots
6,001 to 14,000 feet   210 knots
more than 14,000 feet   230 knots

Wind corrections: Adjust outbound leg to compensate so that *inbound* timing meets the applicable criteria.

CHAPTER VI

The System

So far, you've built your instrument flying environment in layers, first learning to fly an airplane by cockpit references and then adding correcting information from ground navaids of known position. If yours were the only airplane in the world, that's all you'd ever need. So far, the systems you've learned about — and the tasks that go with them — have been centered around your self-contained world in the cockpit. You've even learned the first two elemental tasks in a kind of simplified priority —

> I Aviate
> II Navigate

This describes their functional relationship as well as providing you with a discipline to which you can reasonably adhere in those tense moments when rote rules are the only ideas afloat in a sea of apprehension.

The third and last layer in this onionlike structure of instrument flying is best described as The System, and its role in your flight is encompassed in the third element of the rote rule, which now reads:

> I Aviate
> II Navigate
> III Communicate

and may lead you to think that all that's left to learn is how to talk on the radio; there's a lot more to it than that, though. IFR communication implies the existence of a system; after all, you talk to people for some reason. The reason, as it happens, is that they own the airspace.

The regulations governing instrument flying divide the responsibility this way: you, the pilot, are in charge of the airplane; they, the controllers, are in charge of the *controlled* airspace (which just happens to define almost every piece of air you'd ever use in routine instrument flights). Controlled airspace contains an incredibly dynamic system of assured separation between aircraft operating at reported or radar-observed positions. Effectively, it is a system of mobile reserved buffer zones within the airspace itself; the buffer zones surround each individual airplane and are routed according to the wishes of the pilots, modified by the competing needs of other pilots.

Too many pilots somehow get the impression that air traffic controllers command airplanes. Actually, they couldn't care less what you do with the airplane so long as it conforms to airspace regulations and the need to protect airspace that's been reserved for somebody else. It's up to you to fly the airplane and to *communicate* the need to use certain airspace on your desired course in a timely manner. Now, where the conflicts occasionally arise is in this little loophole: air traffic controllers may not be able to honor your request or you may not be able to honor theirs. He or she may have to offer you an alternate route, a delay, or an airspeed limit in order to fit your moving buffer zone in between all the others. Sometimes, commands that tell you where to steer or what altitude to fly or what airspeed you must maintain sound as if the person on the ground has taken over responsibility. Think about it, though, and you'll very quickly realize that those areas are the only characteristics of airplane performance that a controller can work with in order to deal with competing needs for the same piece of airspace at the same time.

One of the first realities you should understand about the airspace system is that you can *request* anything. You may not get what you request, but as pilot in command, you're allowed to ask them for anything you want. Unless you accept the responsibility for directing your own airplane within the system, you're missing the whole point of air traffic control. And don't mistake the emphasis on this point. "Request" does not mean "demand." If you have advised ATC of a genuine emergency condition, they may well waive some of their own strictures in order to preserve you and your airplane. Unfortunately, there has been a great deal of misunderstanding that has arisen about the relationship between pilots and controllers, and on the one extreme, pilots charge around with a chip on their shoulders, trying to elbow their way through the system. As soon as one of their requests is turned down, they begin to assert themselves to the ATC people, demanding "service."

On the other hand, air traffic controllers sometimes go off the deep end in the other direction, partly because of the way they view the situation (the sky looks rather different depending upon whether you look at it from a radarscope or from a cockpit) or because they get jealous of their control over the airspace. When that happens, their responses to pilot requests can appear unreasonable or simply unrealistic. The fundamental problem is that neither party has a complete picture of what the other is doing at the times when it counts the most. If you could see the picture the controller sees (maybe someday you will) it would probably have a profound influence on the kinds of requests you make. Place him or her in your seat and the reverse situation might well be true.

The system of instrument flying in controlled airspace really places both pilot and controller in the roles of distant partners with the same shared goal: to conclude the flight at its destination in the safest and most expeditious manner possible. During the time you spend as an instrument pilot, you'll find that controllers are human, that they may occasionally err, that you will make mistakes too, and that together, the two of you will be backing each other up in a relationship that is absolutely unique in all professions, for you may never actually meet one another except as voices. If you choose to view every new meeting as a confrontation between you and an obstinate system, chances are that your expectations will be satisfied. If you see your meeting as an encounter with an equal whom you assume to be a professional at the ATC trade, you'll give a little and take a little and probably enjoy life in the IFR environment a lot more. The keys to enjoying your use of The System are an understanding that you are responsible for directing your airplane, a thorough knowledge of the rules for using the airspace, the ability to communicate your requests in a clear manner, and finally and perhaps most important, the patience to view the entire process with a certain detachment. To get your way doesn't mean you've won some contest; it means the airspace was available and ATC was able to handle your request. To be denied doesn't mean you lose; it means that your partner in this slow dance has had to consider some factor that prevents his or her saying "yes" to your request. From your seat, you can't tell which radars are operating and which are down or that a pilot just ahead of you reported a navaid malfunction.

If, over the long term, some procedure seems repeatedly imposed, more often than chance would dictate, and it requires you to make unreasonable deviations from your planned flight, you're justified in making a phone call after you've landed. But please — have your discussions after you've parked. Most of us who have to share the air with you are tired of hearing arm-wrestling matches on the radio between irate pilots and steaming controllers.

Instrument flying rules can be used to fly in good weather or bad. You can file IFR when it's clear in order to gain access to radar-controlled airways and altitudes that provide an extra measure of assurance of separation from other traffic — not a guarantee, but an extra measure of assurance. (One of the best reasons for filing in clear weather is to stay current in procedures; when you're starting out, every flight filed IFR will teach you something new.) However, when the visibility drops below certain minima, the system imposes itself upon you whether you like it or not. (Funny, but pilots and controllers almost never use the proper plural — "minima" — and if you use it, there'll probably be a long pause on the air; most aviation people speak of "minimums.") Also, higher altitudes usually connote high performance, and for that reason, airspace up there is always controlled in order to separate high-speed traffic for which visual methods don't suffice. It is also possible to merge visual flying with the ATC system by flying "VFR on top," in which you use the VFR altitude system and visual separation within the IFR flight plan and with most communication requirements intact.

You communicate with the system when you first call for a weather briefing, when you file your flight plan, when you talk with the tower and departure — all the way to the final "so long" as the line people chock you for the night. The system is incomplete without two-way radio communication between the pilot and controlling facilities. When two-way radio communication fails, you don't always know immediately whether it's your radio or theirs, so it may take a while to establish whether there is, in fact, a breakdown of two-way communication. Many pilots are too eager to use 121.5 MHz, the well-known VHF emergency frequency, to solve any and all problems, including communications failure. Some pilots spend too much time talking on the radio when they should be listening, and very few seem to know all the ways to fly safely without touching the transmit button.

If you are flying IFR inside controlled airspace, no matter what the weather, loss of two-way radio communications requires that you —

1. continue VFR, if the failure occurs while conditions are visual, and land as soon as practicable, or

2. continue IFR according to the radio failure procedures specified in FAR 91.127.

Remember this: if you are in visual conditions when something goes wrong, do not enter conditions where the visibility is less than VFR minimums. You are no longer using the system, and that means you should change to a VFR altitude and land.

The system also assumes that you will navigate along the airway centerline while you are in controlled airspace, or, if you are flying off an airway, that you'll fly in a straight line between the two navaids or fixes defining your direct course. Once you are off airways, you become responsible for terrain clearance, however, since the system defines minimum IFR altitudes only within the airway system and around terminal areas. A rule (FAR 91.119) states that you must select an altitude that clears any obstacle within five *statute* miles (the exceptions to nautical mileage sometimes boggle the mind) by 1,000 feet, or in mountainous areas, by 2,000 feet. (That's one good reason to carry sectionals; there are others.) The same regulation allows you to descend from the minimum en route altitude (MEA) to the published minimum obstacle clearance altitude (MOCA) on a route segment when you estimate you are within 25 *statute* miles of the VOR that defines the route.

Reality is something else again, for it is usually air traffic control that communicates the altitude you must fly based on their situation at the moment when you enter their airspace. If you don't want that altitude, you may request another, and it sometimes helps to state briefly the reason for the request, as the controller may have to adjust other airplanes because of it. You don't *have* to tell him why, but then he doesn't *have* to give you the altitude. Controllers are usually pretty cooperative with altitude requests, however, and even if you don't get immediate approval, they will usually respond with an offer to allow the change after a few more miles or minutes.

Aside from the scattered rules that require you to report certain operations, failures or conditions, all you need as an instrument-rated pilot to operate within the system is

— a flight plan (your airspace reservation request) and
— a clearance (their confirmation of your reservation, with any necessary amendments).

The rules that govern IFR operation in controlled airspace are really very minimal in describing how the operation works; they simply tell you some of what you can and cannot do, and they can be summed up in a scant few paragraphs within Subpart B of FAR 91. The strange aspect to the system is that so little of it is defined as "Flight Rules" and so much of what you really need to know is composed of organizational structure and facilities. So your flight plan and the clearance, your two tickets to ride, involve you with that third layer — communication — and its implied relationship with air traffic control more than they do with any rules.

The trouble is, the printed rules and other publications give you limited access to how ATC really works. In order to learn its basic machinery, a pilot must turn to the *Airman's Information Manual*, which is not a regulation but an advisory publication that describes from the pilot's point of view just how the ATC system operates and what its capabilities and limitations are. It may not mean to, but it severely limits the description of how controllers work — to what criteria and so forth. For that, you need another publication called the *Air Traffic Control Handbook*, which contains buried within its pages the many reasons why controllers have to do the things they do, such as turn you when you don't want to be turned or deny you a descent until you've flown past some fix or other.

The *United States Standard for Terminal Instrument Procedures*, otherwise known as the TERPs, isn't even required reading for an instrument rating, yet it contains the information that visibilities are expressed in statute instead of nautical miles (which is nice to know, since that's the way the regulations define VFR minimums) and provides dimensional data for all the standard approaches and obstacle clearance criteria for a missed approach segment. If you'd like to know the obstacle clearance radii for a circling approach

— within them, they guarantee only 300 feet of clearance, by the way — the TERPs is the only book where you can find that information.

Obviously, part of the system is communication, part of it is defined airspace, very little of it is regulation accessible to pilots, and all of it is spread out into diffuse recesses of various publications and agencies. You can obtain current weather and flight information from Flight Service, file a proper flight plan with the correct (read "legal") alternates and all, preflight the airplane and ensure all the equipment is up to snuff — and yet beyond that, *nothing is certain.* The system is indefinable, changing, fluid. Even after you have received a clearance and are proceeding on your way, that clearance can be *significantly amended at any time.* It's important to understand this aspect of IFR flying, lest you come to envision it as a means to guarantee your arrival. The air traffic control/IFR system is a way of trying to deal with weather and traffic conditions, the best we have right now. It is this limitation to the system that makes it just as important for you to know when to *cancel* IFR or land somewhere unplanned as it is to know how to fly an approach to minimums. It is also important for you to realize very early in your IFR experience that just because you have filed a certain flight and planned for fuel use for that filed plan, you must continue to be aware of changes to that plan and how they may affect your flight. A change in one leg of your plan may be enough to cause your house of cards to tumble and force you to rethink completely the remainder of the flight. FAR 91.23 states that no person may *operate* a civil aircraft in IFR *conditions* unless it carries fuel sufficient to meet the rule's minimums. Since you cannot add fuel in the middle of a flight, that nasty little regulation can trap you. Your fuel may have been sufficient for your planned flight, but it is possible for an amended clearance to render your fuel on board below the legal minimum required for continuing in IFR conditions. You can solve that problem in one of two ways: cancel IFR if you are visual and can land visually somewhere, or change the airport of intended landing. (If the first of the two alternatives applies, you really aren't in a pickle anyway.) It has even happened that pilots awaiting takeoff have received clearances that made it illegal to proceed into the flight.

At first, this can seem downright inconsiderate of them. How can they do this to me? The reason is simple and it's already been stated: they are responsible for airspace; you are responsible for the airplane. It's not the job of ATC to manage your fuel or even to apprise you of an embedded thunderstorm that's astride your airway and five miles from your nose right now. Even though they may see the precip on the radarscope, weather decisions are up to you, and if you want to find out about such a storm, *you have to ask them.* You will get no clearer definition of the divisions of your respective responsibilities than those moments of reflection you spend inside a dark cloud while the precipitation static clogs your com radio and they don't even answer. One comfort: your airspace is reserved just for you — probably because nobody else would want it right now anyway.

To get some idea of the overwhelming proportion of airspace that is controlled to within 1,200 feet of the surface, open a sectional chart or a Jeppesen en route chart. The sectional makes it tough to find, but look for a light blue tinted border — the legend shows a sample under "air space information" — and then try to find areas that are *not* enclosed by such a border. Jeppesen tints *un*controlled airspace with a light blue tint, and they tell me their bill for light blue ink is very low this year; look at the charts and you'll see why. For all practical purposes, I wouldn't worry about uncontrolled versus controlled airspace when it comes to IFR. You should understand that the system will not issue you a clearance in uncontrolled airspace, however. This does not say you can't fly IFR there; it's just that they won't clear you. Remember that clearances are not required in uncontrolled airspace. Most of the uncontrolled off-airway airspace that lies beneath the Continental Control Area (CCA), which starts at 14,500 feet and covers the forty-eight states and Alaska like a huge floating blanket, is found in the western part of the country. These patches of uncontrolled airspace may appear to be tempting, but you'll figure out sooner or later that most of them occur in areas where there's high terrain anyway.

Another sneaky truth is that ATC facilities are not responsible for the navaid system; that falls to the Flight Service Station folks, who monitor each navaid's performance. Controllers will probably know it when a navaid is out, but they are under no particular obligation to tell you something you're supposed to know from the notices to airmen (Notams). In reality, if a navaid has just failed, you'll probably get sick of hearing about it. When a navaid has been out of service for some time and controllers can reasonably expect that you've had time to pick up either a Notam or a Jeppesen weekly avigation revision notice, they're on pretty solid ground if they state that it was your responsibility to know about it.

Normally, air traffic control in this country is conducted in what's called a radar environment. As soon as you hear the words "radar contact," it's your signal that most of the position reporting procedures can be dispensed with. You're foolish if you don't continue to work your calculations and maintain a current estimate to the next fix, because radar always breaks when it's not supposed to. You think they tell you "Our radar's going to bust in 10 minutes, so get ready"? No way. That's why you have to keep up the paperwork.

There are blind spots in certain areas of the country where center radars or more local approach radars can't quite reach, either because of terrain or some other limitation, and if you fly a route often enough, you'll come to know where those blind spots are. But guess what: they're not published. Sure, the ATC folks know where they are, but there's no way to find out about them ahead of time. There are also certain terminals around that may surprise you by lacking radar. They still use time, position reporting, estimates, and paper flight strips to follow you through the approach and for traffic separation. When the controlling facility can't see you, the accuracy of your position reports and estimates becomes critical; it's a time to make absolutely *sure* that was Bogus Intersection you just passed and not Botch. Nonradar terminals are identifiable by a *lack* of a little [R] in brackets next to their names in the summary of the approach/departure control facility frequencies in communication listings and on approach plates. It doesn't exactly leap off the page at you and proclaim, "Hey, watch this one — no radar." In a world of radar, they never fail to come as a surprise.

You'll hear controllers talk about two different kinds of radar: primary and secondary. Primary radar is the form in which radar first appeared. A reflected signal from the target airplane is received and processed for azimuth and distance and displayed on the radarscope without being enhanced in any way and without any data added to the signal. What the controller sees is a kind of smudge on the scope, a bright spot on the phosphor coating of the TV-screenlike presentation. The coating on radar equipment is specially made to hold the image longer than the type on home TV sets, so the bright spot *stays* visible for quite a while. Some controllers in en route centers near busy terminals like to have the older primary radar handy because of that slow image fade characteristic. The targets form a path like a snail's trail and give a very clear picture of how each target is moving with respect to other traffic. If a controller knows he has to merge several converging targets, their respective progress is important information. Primary radar is a primitive form, but it's still useful and that's why it's still around.

Think of a radar site as an electronic searchlight, because radar transmissions are exactly like light except that humans can't see those wavelengths. You've probably seen a radar antenna near an airport, with its emitter horn and huge reflector. It doesn't take much imagination to see the comparison between the filament of a very strong spotlight bulb and the reflector to concentrate the light. As the radar beam travels outward, it strikes solid objects and a small part of its energy is reflected, just as the spotlight would be if you ranged one around a large, darkened room. Primary radar represents a simple, passive reflection of energy.

Secondary radar is something else again, and it's anything but passive. It's as if, as you were swinging your spotlight around, each time you struck particular objects with its beam, the objects were somehow equipped to light up their own neon signs in response; that's the fundamental model of the secondary radar and airborne transponder system. It is formally known as the air traffic control radar beacon system, or ATCRBS. Take a closer look at an ATC radar near an airport and you'll notice a long, narrow antenna atop the huge primary dish; that long antenna is the one for the beacon system. It doesn't have to be as large or as powerful because it operates differently. Instead of simple passive reflections, which obviously send back energy in proportion to the power of the emitter, the beacon-type radar is more like a pulse communication system — two sets of radars talking back and forth. The antenna still helps to determine the azimuth of the target by its position when it receives a response, but beyond that, the differences between secondary and primary radar are total.

The secondary radar transmission is a series of coded pulses, which, when received by the airplane's transponder, is evaluated (after all, there are a lot of radar transmissions all around you) to make sure a response is appropriate. If one *is* appropriate, a carefully timed series of pulses is sent back to the ATCRBS antenna for processing by a computer. This computer and the radar's display system combine to present to the controller very clear, so-called digitized images that include a symbol for the target itself along with a data block that may include identifying numbers and helpful information like altitude and groundspeed.

These two basic forms of radar can be handled in different ways depending upon what type of facility is processing the electronic signal from the antennas. In parts of the country where the volume of traffic makes the radar's job exceedingly difficult, the primary signals are cleaned up on a very sophisticated computer that's able to evaluate each target and determine where it is most likely to be in the next sweep of the antenna. If the prediction and the returning signal match, the computer will more confidently display that target to the controller; it's one way to filter out echo errors and unwanted clutter from the already busy scope. Since secondary radar information is already in the form of digital signals, this computer processing is a matter of course for ATCRBS.

En route centers, which are formally called ARTCCs for air route traffic control centers, have a form of radar called NAS, a completely automated display system that draws the entire picture in symbols and thereby enhances clarity and interpretation. Terminal radar facilities — approach and departure controls, in other words — use a different form of radar display called ARTS III (or the most recent version, ARTS III-A). Terminal radar can combine on a single radarscope both primary type displays and the secondary enhanced displays. Since terminal controllers are more concerned with the merging of traffic into more restricted feeder routes and approach procedures, the ability of the primary presentation to depict trends and flow is valuable to them.

ATC radar systems are designed primarily to do one thing well: find and track aircraft. From a controller's point of view, weather return is only a handicap to that primary job of traffic control, and the radar operates on a frequency that is less adept at defining precipitation than is weather radar. You can expect a controller's radar to depict areas of intense precipitation in general outline, but too many pilots expect more than either the radar or the controller can deliver, such as finding a small pathway through a line squall; that takes weather radar.

Although ATC radars are linked in a limited way to the IBM 9020 computer system, which is the huge, electronic clearing house for flight data, flight plans, and the like, controllers are not zombies who simply sit and stare while machines move numbers around. On some radars in some situations, a target will enter a controller's field of view already tagged with its identifier; however, for the remainder, the controller must spot the naked target where you call yourself in, then he or she must manually tag it with data using a typewriter-like keyboard and an electronic pointer. This operation takes a few seconds, and in high-volume traffic, the distracting task can add seconds to seconds until the job gets pretty busy.

As a backup to computerized flight data, a system of paper strips carries duplicate data on it, and each one must be filled in and moved around through final handoff to the next facility. At the busiest facilities, a second controller may function in the role of flight-strip handler so that the person watching the radar won't be distracted by the task of paperwork. The days of paper strips are probably numbered, and a second independent computer system will probably take over the chore, but paper strips have survived by being simple and reliable; computers break down, but paper strips don't.

What's most irksome about the system is that it's so difficult to find out where the airspace responsibilities lie. Jeppesen charts depict the airspace sectors of responsibility within each en route ARTCC, along with little "telephone boxes" that list the center frequencies in those sectors. These lines of airspace division may *not* indicate anything about radar coverage, however; they do generally indicate where communication responsibility changes; not always, though. There are remote areas of the country where you'll find that below the altitudes at which a lightplane operates, there is no communication with *anyone*. You're supposed to be in two-way contact to be part of the system. These gaps in communication are uncomfortable places to be, at least if you worry about legality. What do you do in an emergency, in icing, in something nobody's dreamed of yet? As you say bye-bye to Albuquerque Center somewhere over the Big Bend area of Texas on your way southeast to San Antonio, they tell you to try Houston Center at a time or fix reference, and it's simply understood that you'll continue within the IFR air traffic control system without communication, maintaining your last clearance. The answer to most situations like this one is to use common sense. What can trap you is unexpected weather in an unexpected communication gap. If you have any doubt that you'll make it through a non-com area, it may be better not to say good-bye to the last center. Yes, it is considered *controlled* airspace: you're on an airway at an IFR altitude. They just don't tell you *how* they control you.

The ATC facilities with responsibility for the greatest volume of airspace are the air route traffic control centers. These centers are closely identified with large cities, although they may be located miles away from their namesakes. The center is housed within a large, air-conditioned, modern building with complete facilities for the working controllers, who operate in shifts twenty-four hours a day.

In the heart of the center is a huge room filled with equipment and radar consoles and usually with some chart carrying weather data for principal airports within eyeshot of every console. The system here is divided into low-altitude and high-altitude structures and also by individual sectors such as the ones identified on the en route charts by frequency boxes. Each individual controller is wired for sound, both by landline to other facilities and by VHF and UHF communications to aircraft. Thus, a center is really a huge octopus of communication and radar links. Both radar and communication is remoted to various locations throughout the sector's airspace to ensure adequate line-of-sight radar and com coverage.

Next in line in the ATC hierarchy is the local facility, which may range in size from New York's Common IFR Room, a giant radar center that dwarfs some of the ARTCCs and handles all approach and departure traffic into metropolitan New York City, through the Los Angeles Tracon, the local facility that receives the "string of pearls" after handoff from L.A. Center and provides those astonishing lines of landing lights stretching outward and upward for miles into the night. Then there's the tower-size approach control, of which there are hundreds, typified by the single controller who may even wear three hats: approach/departure control on a radarscope, tower controller for runway clearances, and ground controller for taxiing. All the ATC facilities are linked to each other by landline and to the FSS network as well. The communication network is established along functional lines, so that a controlling facility is likely to be hooked up to its ARTCC and to adjacent local controlling facilities as well as the FSS for its area.

Hidden within the physical ATC system is a huge computer network that both stores and handles flight data for the IFR activity in controlled airspace. Scheduled flights such as those used by airlines and some corporate operators who fly the same route daily are stored permanently rather than being filed each time the flight departs.

The radars too are joined to computers, which process the primary radar data and turn it into a very crisp display along with tags called alphanumerics with flight data that follows the target along its course. At its most sophisticated, a computer will display on the radarscope not only flight number/identifier but altitude, whether the aircraft is level, climbing, or descending, groundspeed, and terrain warning.

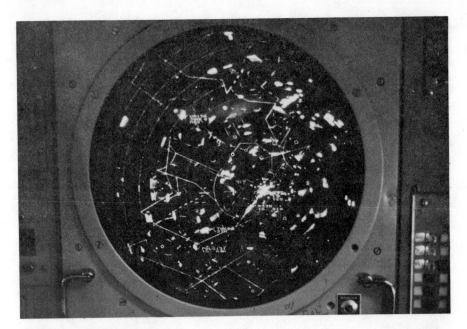

Controller radarscope

The lineup of consoles at a typical ATC center

Alphanumeric scope shows flight number/identifier and altitude.

Alphanumeric readout lists flights under controllers' supervision.

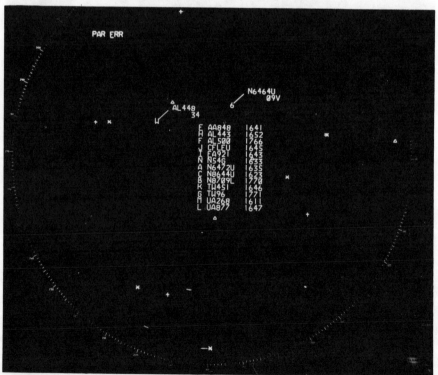

Now that you have a more complete picture of a typical ATC facility, the chore facing the controller takes on more flesh. Controllers have an entire spectrum of people to talk to, not just you. In fact, most of their conversation goes on where you can't even hear it. One of the sneakiest examples of hidden communication is the UHF frequency, where military aircraft conduct their ATC chatter. You may frequently hear a controller talking to an aircraft and you can hear only one side of the conversation; chances are, it'll be the militia on the other end using UHF.

As you progress through the system, the 9020 computer predicts your path and clears the way, figuratively speaking. At the same time, controllers are talking to each other as you are passed from one sector to another and from center to center or approach control. Such handoffs are discussed on open lines between controllers or between the interconnected computers themselves without the need for time-consuming voice communication. But either way, before you hear the words, "Contact Facility Center on one two eight point eight," that next controller is already waiting for you, flight data in hand. You switch to the assigned frequency, and all he wants to hear is who you are, your altitude, and any change in altitude. The longest report you might make, then, would say something like, "Facility Center, Ercoupe four five six seven Papa, five thousand three hundred descending to four thousand." That's all. They know the rest, and if they don't, they'll ask. If any single characteristic marks the new instrument pilot, it is a willingness to tell controllers more than they ever want to know.

When you are handed off between two different facilities, using separate radar units, you may be asked to confirm the altitude readout of your altitude-encoding equipment. Once the controller has established that your encoder is operating within his allowable limits, you probably won't hear about altitude again unless you err from what you've been assigned. They like to check it first, though, rather than assume that what they see is what you're flying.

The air route traffic control centers handle en route traffic control in the airspace covering the entire United States through a system of interlocking regions, each with several sectors within it. Each of these centers is responsible for maintaining traffic separation standards for en route traffic along its portion of the almost 200,000 miles of low-altitude Victor airways.

Standard radar separation is a minimum of 3 miles laterally and 1,000 feet vertically between aircraft within 40 miles of the radar antenna and 5 miles for aircraft beyond 40 miles from the antenna. Altitudes for conflicting east-west traffic are organized according to the time-tested eastbound-odd, westbound-even system.

Within, and usually beneath, the airspace for which a center is responsible may be separate airspace under the control of a local radar traffic facility, such as an approach or departure control. Many pilots erroneously believe that terminal control areas have their own special radar agencies. TCAs are merely an operational airspace definition allied with very busy airline terminals, and they have nothing to do with the structure of the radar-defined airspace.

Because lightplanes operate at low altitudes that may include airspace "owned" by a local agency, you may be flying along at, say, 5,000 feet eastbound and the center will tell you to contact a local approach/departure control and may even give you a new transponder code and certainly a frequency change. You may think that they've mistaken your destination and want you to land, but don't worry; you're just passing through the local agency's airspace, and they'll turn you back over to center when you exit.

Jeppesen and the NOS list communications frequencies inside the fold of each en route chart with appropriate voice calls in boldface type for the airports within the area covered by that chart. Still, there is no chart that will tell you how the airspace is defined and who's in charge of what. It can make quite a difference, sometimes eliminating delays. A friend of mine figured out that in parts of Pennsylvania, his clearance varied with the altitude he filed for. Whenever he filed high, he'd be an hour waiting; lower down, he could get a prompt departure and very little en route delay. Turned out that the higher airspace was owned by New York Center, whereas the lower ones came under the control of local radar facilities, which were less prone to congestion than the larger center with all its immense airspace and eternal traffic problems.

I wish there were some simple way to figure those things out. Most of us have tried everything — calling Flight Service, querying local pilots, anything to abbreviate the procedures. Sometimes nothing helps, and you just have to learn how it works in a particular area. Once you discern a pattern to the routine, it may be worth a phone call to the appropriate ATC facility to inquire about airspace definitions and altitude limitations.

Because IFR flying is so intimately involved with the ATC system, you begin to think of them as one and the same, whereas traffic control is a function of controlled airspace, which has no relation to IFR. IFR is defined by visibility prevailing *inflight*. Hard to believe, but *you* are the one who determines that you're IFR, not a controller. Granted, they may be pretty suspicious of your prevailing visibility if you're flying right through an area where they're painting weather and a flight just ahead of you at your altitude begged for a turn out of there because it was so bad. Because a controller views the world through a radarscope and because the regulations have been written the way they have, there is really no way for ATC to control the imposition of instrument flying rules or to establish through direct evidence that the visibility is anything other than what you say it is. Yes, you could fly at a VFR altitude through cloud and nobody would be the wiser, and there have been cases of pilots who've done exactly that. In one case, it took a long and patient program of record keeping to establish that the pilot in question had, over a period of three years, consistently flown without filing through IFR conditions.

The system is founded on the assumption that we're all sane enough to understand the risk of barging into cloud under visual flying rules, and no, there is no effective means to enforce the IFR rules. In civilized countries, you're allowed to buy an automobile because it's assumed you won't be driving it into crowds of people. That's the way it works with visibility and air traffic control — an honor system, if you like, but one that's based on a reasonable expectation that most of us have an overwhelming self-interest in survival. Still, the number of pilots who are out there, without ratings and with no intention of filing around instrument visibility conditions would probably surprise all of us. (A simple way of determining your inflight visibility has been included in the IFR Notebook at the end of this chapter.)

Whether you are on the ground or airborne, your access to the computer that reserves airspace is through Flight Service. This occasionally gets confusing, because a request for clearance (your "reservation") may be honored if you direct it to a center or to an approach control facility while you're airborne. When that happens, consider it a favor on the part of ATC, and your

A rare moment of relaxation in one of the world's busiest towers: Van Nuys, California

expectation of success will vary across the United States. In California — particularly in the Los Angeles Basin — the prevailing IFR condition is a very limited layer of fog and stratus, usually with clear weather on top. West Coast ATC has adjusted to the reality of the weather pattern and has become used to the so-called pop-up clearance, an abbreviated clearance without the usual complete data from either the airplane or from the people issuing the clearance itself. A clearance to VFR-on-top is a typical use of an instrument rating in California, and the reason for its popularity is that the local ATC facility can issue that sort of limited clearance without going through the full procedure with the nearest center and taking those nice people's time; the clearance is prefiled, and the local facility merely activates it.

The only limit built into the VFR-on-top clearance is customarily an altitude at which you are expected to be VFR; usually, they know enough about the weather up there to be relatively certain about the cloud tops. But if, for some reason, you aren't VFR (a thousand feet above the point where you broke clear of cloud) by the altitude at which they expect you to be visual, you're expected to say so very promptly and then maintain the altitude limit. Proceed any higher and you may encroach upon some other controlling agency's airspace but without the appropriate control.

Similarly, you may depart VFR and arrive at Los Angeles to find that layer of stratus imposing IFR landings at your destination airport. In that case, you would proceed VFR, holding VFR if necessary, while begging a clearance to land IFR from the approach control facility for your destination airport. Although you may never talk to an en route center ATC controller, your flight will be followed by a center's computer.

Another case of abbreviated clearances with local control is known as the tower-to-tower clearance, which is a clearance between two IFR airports (those with approach/departure controls), usually close to one another, such that no en route segment in center airspace is involved. By restricting altitude or route so that you avoid center airspace, you usually get a fairly prompt clearance between the two locally controlled facilities. Essentially, such a flight is a departure procedure leading directly to an arrival procedure, with no intervening en route leg. Even though portions of the flight may not be depicted on a standard instrument departure chart (STAR) or on the approach or arrival charts, the en route portion can be considered a *feeder* to the initial approach fix; at least, that's how you should regard it when you plan such a flight. You'll find nothing in the *AIM* about this local system, yet for certain areas of the country, I'd guess it makes up the numerical majority of IFR operations.

In the case of Los Angeles, there's a published card called the "L.A. card" with preferred routes to use between the airports within the Basin. But take the case of a flight between New York's Westchester County Airport in White Plains and nearby Teterboro Airport in New Jersey — how do you file that one? You can stare at your charts for an hour and find no inspiration there. Call

Flight Service, and the answer turns out to be filing "low-level radar vectors." Be aware of abbreviated clearances when you use such a system for the occasional convenience it affords. When you receive the clearance, it must contain sufficient detail to provide you with an emergency procedure in the event of lost com. Always evaluate any clearance you receive in that light: does this tell me what to do if I *can't talk* to somebody?

After you've planned your flight and are satisfied with the weather information you've gotten, you gain access to the system through Flight Service. Normally, you phone them well ahead of time to file. In some parts of the country, a system called "Fast File" has come into use. It's an automatic telephone answering system that records on tape any flight plan — VFR or IFR — with the proviso that you hang around the phone long enough to allow them to call you back in the event that some portion of the plan won't work or is missing. Periodically, a specialist listens to the tape and enters the flight plans into the computer system. In general, whether you use "Fast File" or telephone a live specialist, you can expect about thirty minutes of processing time before your clearance will be ready.

Although it usually doesn't matter how far ahead you file an IFR plan, there is a limit of about an hour minimum after your proposed departure time, after which they erase your plan from the file. You can call to amend your proposed departure time and keep the plan alive, however.

How and when your clearance is issued may vary with the airport. At smaller airports with little traffic but with either a tower or a Flight Service Station on the field, you may call the ground control frequency for the clearance; unless your instrument chart or airport directory lists otherwise, ground control doubles as an IFR clearance agency at such airports. Larger airports with more frequent airline operations and a heavier volume of clearances usually have a discrete frequency just to handle clearances. It's called a clearance delivery, and it may be listed that way or as "Cpt," which means "clearance prior to taxi." Pre-taxi clearance procedures stipulate that you call the proper frequency not more than ten minutes prior to your planned taxi for departure. Anytime IFR delays exceed five minutes, so-called "gate hold" procedures go into effect to eliminate too many aircraft waiting on taxiways. Unless such procedures are specified, though, you can pick up the clearance anytime. Usually, it's best to call before you start up because you have no way of knowing whether delays will be imposed. It's more considerate to other pilots to take your delays at the tiedown ramp than to clutter the taxiways. Clearance delivery will usually assume you're ready to copy everything the moment you call, so be ready.

Behind the scenes, here's what's been going on: Flight Service specialists have translated your flight plan into the appropriate computer symbology and teletyped it into the ATC computer, where it is filed with all the other flight plans in that center's area of responsibility. At center, it is evaluated and placed in the order it has been received. A route that meets the proper criteria is matched to the request; in many cases, the route will match the request — in other words, "cleared as filed" — but at times, a route different from what you've requested will be constructed and a clearance written.

Printers connected to the computer type out your clearance at the departure airport; at some locations, the clearance is issued over the telephone landline to the local ATC facility to be transmitted to you. Your acceptance of the clearance alerts the system to set aside your departure block of airspace, that you'll soon be on your way. If anything interrupts your plans to depart, you should notify the local ATC facility that issued the clearance to you. Say the engine run-up reveals a bad magneto and you decide to wipe out for the day. Let them know when it happens so they can cancel you promptly; it makes life easier in the airspace around the airport. Remember always that you are reserving airspace, and the sooner you let it go, the sooner it can be released to someone else's use.

This is particularly important in the case of uncontrolled airports with no FSS facility. In such a case, you file and obtain a time block for departure from the Flight Service Station, usually along with a transponder code and a frequency to call as soon as you're airborne, where you'll obtain your full clearance from the appropriate ATC facility on that frequency. The clearance from Flight Service will stipulate that your departure is *void* after a certain clock time (make sure your clocks agree when you call them), and the term "void time" usually defines the departure reservation. They're telling you they've cleared the controlled airspace around your departure for you until that clock time has passed. If you fail to depart prior to that time, you must notify ATC as soon as possible but not later than thirty minutes.

Once you are reported inflight on a departure procedure, your airspace reservations are automatically handled by the appropriate ATC facility as you progress along your cleared route. You don't have to make any further requests for clearances unless you reach a clearance limit and receive no further clearance within three minutes of your limit time. This is confusing, because radar control of traffic is the functional airspace reservation system in today's ATC environment, yet you also operate within a time framework and you don't really know which type of separation is working. You always have a theoretical airspace reservation built into your flight plan: your ETA. However, as soon as you establish yourself in two-way *communication*, whether you're in radar contact or not, that en route time estimate in your flight plan that opened the track for you becomes subordinate to real-time control by ATC. Under radar separation criteria, your flight plan airspace reservation would not be reimposed unless two-way radio communication had failed. So as long as you can talk to ATC, and you are assured that they can hear you, your *most recent clearance* from them has absolute precedence, no matter what you filed or were originally cleared for.

Changes in clearance limits usually occur while you're still en route and the system reconciles your apparent arrival with its prediction for other arrivals. The earlier this occurs, the better, because that's when you have more fuel, more endurance, and more flexibility in your planning. As frequently happens in bad weather, separation standards may reduce the ability of an arrival system to handle large volumes of traffic, and two-way communication is the adjustment that allows the system to manipulate traffic and compensate.

When changes occur — even when you begin a trip VFR and realize you'll have to file IFR in order to complete it, a process called "air filing" — you must communicate with the system in order to determine what airspace you want reserved under the altered conditions. If you are *already talking to an ATC facility*, they become the appropriate agency to whom you communicate your request. The words that ordinarily cue such a response are "What are your intentions?" The controller is literally inviting you to request another chance at the same airspace or use of airspace leading to a different destination. *If you have not yet established contact* with an ATC facility and you need to change your request or file for the first time, you should gain access to the system through the nearest Flight Service Station.

What's different about air filing is that you're already en route and therefore have no departure segment. Your request for airspace should ask for a route to the nearest appropriate navaid or airway to get yourself established in the system in the most direct manner. In order to fulfill the need for some airspace reservation in the event of loss of communication, you must estimate the time between your present position and the destination such that the estimated arrival will allow reserved airspace for a no-communication arrival at the proper time.

Two things change when you operate within the system under a VFR-on-top clearance: traffic separation and altitude. VFR-on-top is not authorized by ATC in Positive Control areas, which is all of the airspace from 18,000 feet to FL600. Although the system expects you to behave under VFR-on-top as if you were an IFR flight in most respects, such as position reporting (in a nonradar environment) and other required reports, it may not provide you with traffic advisories, since the VFR-on-top clearance allows you to pick any altitude with VFR conditions above the published MEAs or appropriate IFR altitude. Above 3,000 feet AGL, moreover, you must use the plus-500-foot altitude system for VFR flying and vacate the IFR altitudes, which operate on thousand-foot levels. As you might expect, the system requires that you adhere to the cleared route and destination; any change, in fact, must be requested. What may surprise you is that your filed airspeed is also an element of the flight plan to which you are expected to adhere, unless you advise ATC differently. It's considered a personal foul, therefore, to decide after you're under way that you'd rather throttle back and ride a tailwind. Alter your true airspeed from whatever you've filed and you must advise ATC; they consider it a change in the flight plan. (With radar, it's been my experience that the only times they really care is when there's a speed *restriction* or when the discrepancy between filed airspeed and your actual apparent airspeed on the scope is significant.) Understanding this point is pivotal to the functional distinction between how you view the world, as the pilot, and how the ATC controller sees you within his or her context.

Air traffic control radar displays are a ground-map world. They show actual course lines of aircraft to the controller. It means that no controller is precisely aware of your dynamic navigational situation — which means the wind at your altitude, really — even though the computer may be aware of the forecast wind. Most controllers very quickly gain a rough estimate of the direction and velocity of the wind at various altitudes, and they may occasionally ask you your heading in order to help them arrive at such an estimate. All they see is your course, so by comparing that to your reported heading, they can come up with a wind estimate, rather like navigation but in reverse.

Because controllers see only ground course and groundspeed, they must rely on you to give them accurate airspeed and heading information. There's no rule that says your airspeed indicator must be correct; however, you'll only be giving yourself problems if you fly in the system with an ASI that's off by some significant number. Once you are being controlled on radar vectors instead of flying under your own navigation, therefore, the controller is essentially doing the navigating for you and must use an estimate of the wind to assign headings. You should expect corrections as the effect of the wind changes and your actual course line becomes apparent on the scope.

When it comes to airspeed, real conflicts can arise. In any situation wherein a controller issues an airspeed clearance that you can't meet, you must say so immediately. It is theoretically possible for a controller to issue a speed control correction that could take the airplane into the yellow airspeed arc or below its stall speed. There is one case where the system can almost seem unfair, although it takes a high-performance airplane to get into such a pickle. The rule says that below 10,000 feet, no person may operate an aircraft at an indicated airspeed of more than 250 knots (FAR 91.70). On a descent, it is up to the *pilot* to ensure that the airplane adheres to speed adjustments within ten knots, but added to that is the sole responsibility to adhere to 91.70 and reduce speed below 250 knots at 10,000 feet. Don't ask me why, but the "Administrator" is the only authority who can allow you to exceed 250 knots below 10,000 feet. What's really strange is that any ATC facility *can* authorize you to exceed the speed limits in the airport traffic area. Confusing? You bet. But there's more. If you're in a TCA, you revert back to the former situation! Again, you become responsible for operating at less than 250 knots indicated, and ATC cannot impose a clearance to do otherwise.

In other words, it is possible, within the system, for ATC to issue you a clearance with which you cannot comply or which would place you in an illegal position if you did.

You got it right.

You *can* get a bad clearance, and one of the trickiest single elements of the ATC system is to stay alert to such a trap. Air traffic control does not purposely set out to ensnare you, believe me, no matter how much it may seem that way at times. These conflicts between their world and ours, with us in control of the airplane and them in control of the air, come to light with the greatest clarity when something goes wrong, and it is therefore worth your life to detect the flaw before it leads you into trouble, whether it's merely a legal point or actually dangerous.

Where the conflict imposes the most potential danger is in traffic separation, descents, and maneuvering anywhere near terrain, where responsibility for terrain clearance may get confused and you may be so distracted by new vectors and anticipated approach clearances that you may fail to separate the airplane from the system. Of particular danger to lightplane pilots is any change in procedures during climb portions of a departure. Always evaluate such situations in terms of your airplane's ability to deliver. Controllers have become increasingly sensitive to terrain clearance, and are usually very careful to check with lightplane pilots whose rate of climb may be limited in areas where terrain defines minimum acceptable climb rates.

All this pales before the single situation that seems to give pilots the most trouble: vectors off published routes.

Part of the problem is caused by the way the system has constructed one set of minimum altitudes for the controller and a different set for the pilot. A controller is guided by minimum vectoring altitudes, whereas the pilot must rely on the sectional charts as the only assurance of off-route terrain clearance. It is simply unrealistic to expect a pilot to fly an airplane and maintain some precise notion of position on both an instrument and a sectional chart. Instrument charts publish minimum safe altitudes, which are considered to be emergency-use-only information on approach procedure charts with VOR or NDB facilities in place and operating.

Your problem is that you are responsible for ensuring adequate terrain clearance and for questioning any vector or clearance that may be unsafe. On an approach controller's scope are marked sectors defined by obstacles within the radar service area. Each sector has a minimum vectoring altitude that assures 1,000 feet clearance from the highest obstacle within three miles; in mountainous areas, that goes up to 2,000 feet, except in cases where approach procedures require less. These minimum vectoring altitude charts bear no resemblance to the minimum safe altitude depiction on the approach chart. The book further states that the minimum vectoring altitude may very well be lower than published MEAs or MOCAs.

How can you protect yourself in a system like this? First of all, be alert to the difference between vectors and your own navigation. In the unlikely event that an approach radar facility might vector you for a short distance and then return you to "resume normal navigation," such as a clearance direct to a navaid from your present position, there is an implied return of primary altitude responsibility to the cockpit. Thus, if the controller assigns you a *heading* to fly and says the magic words "radar vectors," it may be safe to accept an altitude that's lower than a published MEA or MOCA or even the minimum safe altitudes as shown on the approach plate chart. In any situation where you are navigating on anything other than heading assigned by ATC, you are navigating on your own and should return to the published MEAs.

In the past, vectors off published routes and leading to an approach navaid have caused confusion to some pilots because they misinterpreted the words "cleared for the approach." When you hear those words, you have a tendency to bury yourself in the approach procedure, which may lure you to lower altitudes than ATC intends. Remember that an approach clearance is not a license to descend to a published approach altitude *until you are established* on either a published route (a feeder route, say) or the initial segment of the approach procedure. And in both cases, you'll have a published altitude to hang your hat on.

Obviously, if every pilot nagged every approach controller who ever issued a vector, questioning every altitude assignment just because the MVA is not available to the pilot and the assigned altitude is lower than the pilot's published minimum, the system would bog down. Some reasonable compromise between safety and efficiency has to be struck, and fortunately there's an easy way out of the situation. Planning.

That's right, planning. Having a reasonably accurate picture of the terrain around your destination is the single best measure you can take to ensure that serious mistakes aren't made on altitude assignments. Jeppesen's area charts for mountainous areas now include topographical maps, that company's contribution to easing the terrain and vector problem. Some pilots don't bother with the area charts because it's one more piece of paper in a wad of en routes and approach plates, but the area charts are much handier than a sectional, and from them, you can get a good general picture of the height of terrain over which you'll be operating. Simply look at the chart in an overall scan to determine the highest points around the destination, then pick your most likely route into the terminal area. Notice where the terrain lies with respect to the route and you'll be sensitive to any turns in that direction. If you question a turn in the direction of terrain coupled with any altitude change or assignment, any reasonable controller will probably respect you the more for asking.

You'll find after some time with the system that it's usually how you ask rather than what you ask that makes the difference. Communication with the

ATC system is quite properly the third ranked chore for you in the cockpit, but that's with respect to the *rate* of your workload, not the importance of the idea. It's just as important to be skilled at communication as it is to be a precise attitude pilot and navigator.

Communication has changed incredibly rapidly within recent years, and the change has been for the better. The more modern radio systems provide you with a clarity of communication that your colleagues of an earlier generation only dreamed of. Where they struggled to get a simple message to the ground ("Words twice, I say again, I spell . . ."), communication with ATC is now as close to a conversation as it can get. Procedural words still have their place, but brevity is to be valued more than knowledge of when to say "break, break" or some other radio jargon. Having a clear idea of what you want and communicating that need with equal clarity in the minimum number of words is the mark of a good communicator, and that's what you should strive for. Whether you say "affirmative" or "yes" doesn't really matter that much anymore except at moments when communication becomes difficult and procedure words aid clarity.

So there is a flight director's style when it comes to talking too. You would not yank a yoke over into a steep bank without good reason, so why push a transmit button before you know where your mouth is going? Someone once said that you could fly coast to coast in perfect safety in this country and never talk to anyone, just *listen*. There's information all around you — ATIS recordings, automatic weather broadcasts, Flight Service transmissions to other aircraft — and if you consider it part of your job to use the *listening* half of the communication system to the utmost, it will alter the way you look at your communication avionics. It's not a device for talk; it's a means to *receive* information. Losing your transmitter is hardly a serious matter in today's ATC environment — so long as you can listen. Don't be one of those pilots who cranks in a new frequency, and almost before it's tuned, keys the mike button. Talk as a flight director would: slow down, tune, listen, think, and then . . . just maybe . . . talk. Unless your sneakers have caught fire, there is absolutely nothing so urgent that it won't wait a few seconds.

Look at it this way: you may be hurtling through the air at hundreds of miles an hour and thousands of feet up; it's enough to fill you with a sense of urgency. Your position with respect to reference navaids may be nudging the speed of sound — *but your rate of change with respect to the ATC system is always zero.* They're way ahead of you.

If it were up to me, I'd add to the priorities

Aviate
Navigate
Communicate

IFR Notebook

Airspace

VFR MINIMUM VISIBILITY AND DISTANCE FROM CLOUDS IN STATUTE MILES

Altitude	Uncontrolled Airspace		Controlled Airspace	
	Flight Visibility	Distance From Clouds	Flight Visibility	Distance From Clouds
1200' or less above the surface, regardless of MSL Altitude	1 statute mile	Clear of clouds	3 statute miles	500' below 1000' above 2000' horizontal
More than 1200' above the surface, but less than 10,000' MSL	1 statute mile	500' below 1000' above 2000' horizontal	3 statute miles	500' below 1000' above 2000' horizontal
More than 1200' above the surface and at or above 10,000' MSL	5 statute miles	1000' below 1000' above 1 statute mile horizontal	5 statute miles	1000' below 1000' above 1 statute mile horizontal

What this chart means is that if you can't see at least as far as the minimums, you must at least have an instrument rating and the required equipment for instrument flying; in controlled airspace you also need a clearance.

How to Estimate Inflight Visibility

The approximate visibility in miles will equal the number of thousands of feet above the surface when the surface is just visible over the nose of the airplane.

A more accurate method of determining visibility using the above scheme of cockpit cutoff angle: Determine on the ground the tangent of the cockpit cutoff angle:

$$\frac{\text{height of the pilot's eyes above the ground (ft.)}}{\text{measured cutoff point in front of airplane (ft.)}} = \text{tangent}$$

This tangent is the ratio of height (altitude) to forward visibility in *feet*.

If you determine, using an estimate of inflight visibility or distance from clouds, that conditions are below VFR minimums, you are already in the system! Other traffic could be anywhere, not just at IFR altitudes. It could be climbing or descending between IFR altitudes. *Maintain VFR and get a clearance.*

Controlled and Uncontrolled Airspace VFR Altitudes and Flight Levels			
If your magnetic course (ground track) is	More than 3000' above the surface but below 18,000' MSL fly	Above 18,000' MSL to FL 290 (except within Positive Control Area) fly	Above FL 290 (except within Positive Control Area) fly 4000' intervals
0° to 179°	Odd thousands, MSL, plus 500' (3500, 5500, 7500, etc)	Odd Flight Levels plus 500' (FL 195, 215, 235, etc)	Beginning at FL 300 (FL 300, 340, 380, etc)
180° to 359°	Even thousands, MSL, plus 500' (4500, 6500, 8500, etc)	Even Flight Levels plus 500' (FL 185, FL 205, 225, etc)	Beginning at FL 320 (FL 320, 360, 400, etc)
Uncontrolled Airspace—IFR Altitudes and Flight Levels			
If your magnetic course (ground track) is	Below 18,000' MSL fly	At or above 18,000' MSL but below FL 290, fly	At or above FL 290, fly 4000' intervals
0° to 179°	Odd thousands, MSL, (3000, 5000, 7000, etc)	Odd Flight Levels, FL 190, 210, 230, etc)	Beginning at FL 290, (FL 290, 330, 370, etc)
180° to 359°	Even thousands, MSL, (2000, 4000, 6000, etc)	Even Flight Levels (FL 180, 200, 220, etc)	Beginning at FL 310, (FL 310, 350, 390, etc)

You may not fly into: Prohibited Areas, under any circumstances. Under certain conditions, when at an ATC assigned altitude, you may fly into restricted airspace, such as Restricted Areas, Warning Areas, Alert Areas, or Military Operations Areas (MOAs) with an IFR clearance.

Airport traffic area: five statute miles in radius from the center of an airport within which a tower is operating, from the surface up to, but not including, 3,000 feet above the airport.

Airport advisory area: five statute miles in radius from the center of an airport with an operational Flight Service Station but *no* tower. There is no altitude definition because an airport advisory area serves a communication/FSS function only.

Control zone: normally five statute miles in radius, with extensions for approach and departure paths and extending upward to the base of the Continental Control Area at 14,500 feet. A control zone becomes effective only when the weather at an airport is less than VFR minimums. This tells you two things: a qualified weather observer must be located there; an airport without a control zone delineated around it has no weather observation anytime. It therefore *cannot* serve as an *alternate* for flight planning purposes. Knowing this makes it easy to spot qualified alternate airports near your destination during your flight planning; check the en route charts.

Transition areas: designated areas connecting the terminal and en route portions of the controlled airspace system, extending upward from 700 feet AGL when associated with an airport and from 1,200 feet upward when associated with a route structure and terminating at the base of the overlying controlled airspace.

Airway: airspace four nautical miles on either side of the centerline connecting reference navaids and from 1,200 feet AGL (or as designated) to 17,999 feet, the base of the Positive Control Area; beyond 51 NM from a navaid defining the airway, the airspace is defined as an angular divergence of 4.5 degrees.

Continental Control Area: airspace above 1,500 feet AGL and from 14,500 feet to eternity — there is no upper limit defined.

Positive Control Area: all airspace between 18,000 feet and flight level (FL) 600 with certain exceptions; in Alaska, excluding airspace at less than 1,500 feet AGL

Control over airspace is not limited to the above listed areas. Air Defense Identification Zones (ADIZ) are an example of a military controlled airspace, and the North Atlantic Track is an example of controlled airspace in international skies.

ATC Light Signals

	Surface	*Inflight*
Steady green	clear for takeoff	clear to land
Flashing green	clear to taxi	return for landing
Steady red	stop	give way to other aircraft and continue circling
Flashing red	taxi clear of runway	unsafe to land
Flashing white	return to start point	does not apply

Alternating red/green — exercise extreme caution

Acknowledge: day — move ailerons or rudder

night — blink lights

Aircraft Speed Limits

Below 10,000 feet — 250 knots *indicated*

Within a TCA — 250 knots *indicated*

In an airport traffic area (except one within a Terminal Control Area) —

reciprocating engine   156 knots *indicated*

turbine powered   200 knots *indicated*

In airspace underlying   200 knots *indicated*

a TCA or in a TCA

VFR corridor —

If safe operation requires an airspeed higher than those listed, aircraft should be operated at the minimum safe airspeed. ATC may also approve operations up to 250 knots indicated.

Altimeter Settings

Below 18,000 feet — to nearest station within 100 NM

Above or at 18,000 feet — to 29.92

Special VFR at Airports

Visibility at least one statute mile

Clear of clouds

VFR at Airports in Controlled Airspace (or, Control Zones)

Visibility at least three statute miles

Ceiling 1,000 feet or better, unless classified as thin or partial

VFR at Airports in Uncontrolled Airspace

Visibility at least one mile during takeoff and landing, as determined by *you.*

Reports Required by ATC

Any unforecast or hazardous weather, even if the weather is *better* than forecast.

Equipment failure that impairs communication, navigation, or aircraft control.

Departing an assigned altitude; any change in altitude when VFR-on-top.

Points named in a direct flight.

Any change in true airspeed of more than 10 knots or 5 percent, whichever is greater, from that filed in the flight plan.

Any missed approach, including one initiated prior to the missed approach point; plus intentions and reasons for the miss.

Unable to climb/descend at a rate of at least 500 fpm.

"Unable to comply" with any clearance issued for reasons of performance limitations, weather conditions, or equipment.

When in a nonradar environment, any deviation of more than 3 minutes from a reported estimated time of arrival (ETA).

When, in a nonradar environment, entering or leaving a hold, including altitude.

Final approach fix (FAF) inbound, when in a nonradar environment.

At a required reporting point as marked on a chart with this symbol ▲, when in a nonradar environment.

At a reporting point marked on the chart with this symbol ₋, when requested by a controller.

When to Note Time References for Required Position Reports

At a VOR, at the time of the "to-from" indicator reversal.

At an NDB, at a time when the indicator arrow makes a complete reversal.

At a beacon, the time midpoint between when the aural signal is first received and when the aural signal is lost.

Elements of a Position Report

Identification
Position (reference a published fix)
Time at that position (see preceding paragraph)
Altitude
ETA, name of next reporting point
Name only, next succeeding reporting point
Remarks (not required or encouraged)

Minimum Altitudes for IFR Operations

Unless necessary for takeoff or landing, no person may operate an airplane under IFR at less than:

— the published minimum en route altitude (MEA)

— in a designated mountainous area (see Jeppesen US-3 or equivalent), 2,000 feet above the highest obstacle within a horizontal distance of five statute miles from the course line to be flown

— in any other case, 1,000 feet above the highest obstacle within five miles of the course to be flown

— within 25 *statute* miles (that's about 22 nautical miles DME without a slant range correction) of a VOR, the published minimum obstacle clearance altitude (MOCA)

Winds

National Weather Service	reference to *true* north/knots
ATC	reference *magnetic* north/knots

Visibility

NWS and ATC: in *statute* miles

Do not squawk or tune through transponder codes:
0000, under any circumstances, unless you are a military interceptor
7500, 7600, or 7700 unless necessary
"ident" mode, except when requested by ATC

Do not communicate with FSS facilities on any frequency during approximately :00 to :15 minutes past the hour; that time period coincides with scheduled weather broadcasts.

Turbulence Reporting Criteria

Light turbulence: slight, erratic changes in attitude; slight strain against seat belts, unsecured objects displaced slightly (term "light chop" used if rapid and rhythmic bumps, like a "washboard effect," are encountered)

Moderate turbulence: similar to light turbulence but more intense peaks, variation in indicated airspeed but no control problem; definite strain against seat belts and displacement of loose objects ("moderate chop" for rapid, rhythmic jolts, more severe than "light chop")

Severe turbulence: large changes in altitude and attitude, with large variations in indicated airspeed; aircraft occasionally out of control momentarily; occupants forced violently against straps and loose objects tossed about

Extreme turbulence: aircraft practically impossible to control

Duration:
Occasional: less than 1/3 of the time
Intermittent: between 1/3 and 2/3 of the time
Continuous: more than 2/3

Elements of report: location
                         time (GMT)
                         intensity
                         conditions (cloud, etc.)
                         altitude
                         type aircraft
                         duration

Icing Report Criteria

trace   perceptible, accumulation slightly more rapid than sublimation (evaporation)

light   rate of accumulation could be operational problem if condition continues for over an hour; deice equipment sufficient to remove or prevent

moderate even short encounters become hazardous, use of deicing equipment necessary

severe   deicing equipment unable to cope with accumulation

Elements: identification, location, time (GMT), intensity (see above), type (rime, clear or mixed), altitude, aircraft type, indicated airspeed.

CHAPTER VII

Planning and Organizing a Flight

IFR flying is planned backward. You pick a destination airport, take a look at the approaches into the airport, and then establish an en route segment that will terminate at a navaid or fix that leads to the primary approach for the winds you expect to encounter on arrival. Once all that is done, you can plan a departure, one that will lead you into the en route airways you want.

Other considerations now impose themselves: aircraft performance is one, the weather is another. Aircraft performance stays pretty much the same except for seasonal changes in air temperature that affect climb, cruise, and runway performance. Weather changes all the time, even during the flight itself, so you never really stop planning for weather effects.

The object of your flight plan is merely to get your shoe in the door, so a flight plan is not necessarily your plan for the flight. Your route, altitude, and alternate airport may change drastically during a flight, so don't think of the flight plan as the sole end of planning. Once the flight plan has gotten you into the computer and delivered a clearance, the rest of the game is open free-style, and in light traffic, given a willing controller, you may be able to negotiate near miracles of expediting. Sometimes it's even the person on the ground who suggests a change in actual routing! So get rid of the notion that you will construct some sort of iron-clad arrangement and then adhere to it. A flight plan is for clearances, nothing more; it mainly meets certain legal obligations. Don't take this so literally that you throw your flight plan away, however; you'll be in a real bind if ATC clears you for "flight plan route."

Start with the fundamental principle that you may file IFR to any airport (and from any airport) in the forty-eight states. Even airports with no published instrument approaches have a legal weather minimum: it's called VFR, and if you have VFR conditions predicted from the MEA downward, you can file to that airport. The rule states only that you can't shoot an approach under instrument conditions into an airport with no published approach.

You may also wish to take into consideration the published low-altitude *preferred routes,* which are published by the National Ocean Survey in the back pages of their *Airport/Facility* directories and by Jeppesen as part of an en route chart subscription. Preferred routes are systems of airways between selected cities, sometimes operating in one direction only between these terminals. If your flight takes you along the airway structure that includes a preferred-route city pair, it may be to your advantage to incorporate the route into your planning.

By the time you have a few hours of VFR under your belt in the airplane you plan to use, you will have a firm idea of its normal cruise fuel consumption and its true airspeed. You'll also get some idea of how the airplane performs with various loads, from a minimum of just you and your luggage to a maximum of gross weight. Planning for an IFR flight is better based on real performance than on the published figures in the operating manual, since the book's figures may be optimistic. Planning considerations for the airplane's actual performance therefore begin many weeks before you commence flying IFR. What will vary most with respect to your airplane's performance will be true airspeed with altitude change. Most airplane flight manuals give you a graph of airspeed for various power settings against altitude, and a normally aspirated engine will have a characteristic *optimum* altitude at which an airplane delivers its greatest efficiency. In general, the altitude at which full-throttle airspeed meets the 75 percent power airspeed reference line is the altitude at which you should try to file. Weather and terrain may affect your decision on this point, but there is no good reason for operating an airplane at less than its maximum airspeed for a given horsepower. Look for that bulge in the graph, and you'll usually find it somewhere between 5,000 and 8,000 feet.

Do most of your route planning the night before a flight based on the weather outlook for the next day. Get the charts out and spread them across the floor to get a general picture of the route structure from origin to destination. Then get a general mileage summary for the most likely route so that you can determine how many fuel stops are likely, if any. It's very rare that you'll encounter long stretches of IFR weather, no matter how long your trip will be. On the average, a three-day trip across the United States in the summer may place you in IFR weather for only a few hours' total time, and for that reason alone, there's almost no situation that ought to cause you to stretch your fuel on an IFR flight. On long flights of that sort, I like to plan a stop just before entering the IFR segment, and I usually hunt for an airport that has a Flight Service Station on the field. Nothing can substitute for an in-person briefing, particularly if the FSS has a weather radar repeater in the office. To me, it's better to begin a trip IFR and proceed into VFR weather than to depart VFR and land on instruments. Save the easy part for when you're tired. Naturally, you can't always do that, but when you have the choice, always land prior to the bad-weather leg. It's a chance to get organized, to refuel, and to give yourself a better opportunity to weigh the decision about proceeding farther.

On the night before a flight, you can usually get a fairly accurate idea of where the stops will be, within 100 NM or so, and rough out a route for the entire flight, including the return leg, so that all you have to do in the morning is make minor adjustments according to the final weather picture. Preliminary planning also reveals a few likely candidates for alternate airports. Since alternates are really valuable only if they have a good viable approach aid, you can eyeball the Jepp en route chart — the airports with instrument approaches are spelled in capital letters — and then take a peek at each candidate's approach charts to see which offers the lowest escape route.

The destination airport approach charts tell you a lot about how tomorrow's weather will affect the chances of completing the flight too. In a case where several airports serve the destination city, one or more can be expected to offer a low, precision approach — and also the potential for traffic delays. An outlying airport may take you closer to your ground destination, and if the weather looks a little easier, an airport with a nonprecision approach might be a good bet. There is no real advantage to an ILS approach in terms of your staying current — all the book says is that you have to shoot six *approaches* within six months; it doesn't say what kind, so although currency considerations in IFR planning may urge you to shoot an approach, they don't push you to a particular airport in easy weather.

Destination charts show one very important planning component: the transition from en route to arrival. In order to plan the en route portion of the flight, you must first determine the likely approach in use. Typically, an airport will use its most precise approach as often as possible, barring adverse winds or noise-reduction considerations. A check of the ILS charts for that airport will tell you where the *feeder routes* to the approach have been designed. Obviously, your flight is going to be cleared to one of those, so you may as well plan for it at the front end.

The destination charts also describe an important component of planning that most pilots ignore: the missed approach. The missed approach is considered an approach segment, and if your airplane isn't equipped to fly it, you may have to take that into consideration in your planning. It's mighty embarrassing to realize as you're missing an approach and preparing to enter a hold southeast at the outer compass locator that an ADF wasn't part of the option list on this airplane.

Now you know where the approach will probably commence, and you've eyeballed the other less likely approaches so that tomorrow, if things change, you'll have scanned them. Next, work your way backward along the airway route structure to the departure airport and see which of the Victor airways describes a fairly straight line between the two. Now ask yourself these two questions:

1. Is there a preferred route anywhere in the route structure?
2. Are any navaids out of service on the airway route?

With a Jeppesen subscription, both questions can be answered by checking the front of the book for "VOR Shutdowns/Substitute Routes" and the "Avigation Revision Notice." Preferred routes between various city pairs are also up front in Jeppesen's service. The NOS publishes the Preferred Routes in the *Airport/Facility Directory* subscription and the shutdowns are published by the FAA as Notams.

Notams come in a variety of flavors and packages, and the whole system has gotten more confusing than it really needs to be. There are two ways the FAA disseminates Notams. One is the Notices to Airmen (Class II) publication; the other is the National Notice to Airmen Service A system, which is a link-up of teletypes and a central computer in Kansas City. There is a categorization of Notam information as well:

Notam-D: information that could affect your decision to make a flight; D-type information includes airports closed, navaids down, and the like — the important stuff.

Notam-L: the "L" stands for "local" and that's the kind of information these contain; advice about taxiway potholes, men and equipment, feeding the pigeons, and the like.

FDC Notams: although these come out of the Flight Data Center in Washington (hence the initials), they're available through the Kansas City Computer Center and accessed by teletype. FDC-type Notams are the most important, since they contain changes that are *regulatory:* alterations to charts, instrument approaches, or restrictions to flight.

Ds are "oughta know," Ls are "nice to know," and as far as FDCs go, you "better know" those. If they are seven days old or available in time for publication, you'll find the Notams in the Class II publication.

Now this is important: *unless you tell them otherwise, an FSS will assume you know the published printed Notams.* The Class II publication contains all Notam Ds that were available in time, a few Ls that someone thought were important enough to include, and every FDC Notam they could get their hands on. The book contains a number system so you can tell at which point you have to begin updating. When you call for an FSS briefing, therefore, be sure you ask for the Class II Notams covering both your route (navaids) and your destination and alternate (airports), along with updates on the Ds and FDCs.

The unpublished D and FDC Notams are available through the Kansas City file and are automatically sent out. Any FSS can also query the computer for a specific Notam. L-type Notams are filed only at the local FSS having jurisdiction over the facility concerned. To get hold of those Ls that are not listed in the Class II book, you have to inquire directly of the FSS responsible, not Kansas City. By now, all this is probably as clear as options trading. Try to make sense of it and do the best you can with the ones that are important. Just asking for the Notams is a big step forward.

Take a closer look at the en route airways for the typical MEAs you'll be encountering. Can your airplane hack it? There are points in the western United States where a light single-engine airplane simply can't meet the MEA safely; oh, you might make it to the 15,000-foot MEA between Hanksville, Utah, and Montrose, Colorado, but could you maintain a 500-fpm climb? What would you do in unforecast ice? What will the true airspeed be at that altitude? You may very well discover that a light airplane simply can't handle some routes if the outlook indicates that you can expect clouds and potential icing. Light airplanes are altitude-limited, something their high-performance siblings sneer at. Is the flight impossible? Of course not.

Get out the sectionals. This may be the perfect time for a composite flight plan. Many of those MEAs may have less to do with terrain than they do with navaid reception. It may be that one portion of your flight should be flown under VFR *below* the weather. Below-the-weather VFR flying is the little airplane's trump card, particularly when the weather is of the kind that's localized (thunderstorms, spotty showers). Just because you have an instrument rating is no reason to be a slave to IFR flight in controlled airspace. A perfect example of an area where a composite flight plan can save the day is a late-afternoon departure in the summertime from Phoenix, Arizona, headed west to Los Angeles; another is Florida, where isolated thunderstorms are the norm. When the sky is littered with anvil clouds, file for an airway and you'll almost assuredly hit one of them right in the middle of its angriest dark spot. You'd be absolutely justified in concluding that the little airplane is probably better off than a turboprop that tries to poke its way through at the middle altitudes. Go VFR, stay low, and aim for the wide open spaces between the storms. It's always amazing to me how many pilots file IFR *only* because the weather briefing sounds bad.

Another good reason to use a VFR segment in your planning is because it may eliminate a delay. This is particularly true for an uncontrolled field near a busy controlled airspace. In order to protect your departure airspace, ATC may have to wait for an opening. If you file for a VFR departure, you're telling them you can handle traffic separation visually until you're well into the en route altitudes and far from congestion. That may eliminate the delay at a stroke.

Quite often, you may find that there's no airway between two navaids along the route you want to fly. Can you fly direct between the two VORs? To find out, simply determine the class of the VOR and the altitude at which you'll be flying. Usually, if the measured off-route distance includes the service volume of both navaids, you'll get what you want. But at what altitude should you fly? Get out the sectional again and check terrain within *five* miles on either side of your route. Add 1,000 feet to that and round it off for east- or westbound headings; add 2,000 feet in designated mountainous areas (they're in the *AIM* Part I along with the ADIZs). For some reason, students are taught by some instructors to file for the highest MEA they'll encounter along a route; I don't know how that got started, but it must have been a flatlander's idea. Try that in the mountains and you'll find ATC sticking you up there for the *entire* flight. There is nothing wrong with filing the altitude you want for the initial part of the IFR flight and simply climbing later on to meet the higher MEA; with fuel burned off, it may well be an easier climb, for one thing, and for another, you'll tire a lot less at the lower altitudes.

By the time you're ready for bed, you've had an extensive look at the system for the route you'll be flying, you have a fair idea of what the weather will be tomorrow morning from the outlook, and the rough time-distance computations even provide you with a guesstimate of your departure time. In the morning, you'll need only to refine this picture. Estimating the next day's weather may also make it easier for you to put off a trip that's better delayed. You can call ahead the night before and cancel; then everybody can relax. If you wait until it's time to leave and suddenly face some awful weather, you may pressure yourself into making the trip. Part of IFR planning is to stay ahead of yourself as well as the mechanics of flying. That means understanding and avoiding situations that may tend to pressure you into doing things you really don't want to do.

Don't finish your planning until you've at least glanced at the return leg of your flight, if it's to be flown at the end of the same day. Even if it simply tells you that the return leg is an exact reverse of the trip out, the time spent is worthwhile. The hidden trick in return legs is the MEA eastbound that just matches your airplane's altitude limit; you'd be expected to fly westbound on the return 1,000 feet higher — can you handle it? Sometimes it depends on where the high portion of the route lies; toward the end of a trip, you might manage with decent weather. Don't depend on ATC to give you the MEA just because your airplane can't make it any higher. Remember the division of responsibility. It's *your* airplane, and you're expected to flight plan according to its performance. It's their airspace, and they don't have to move other traffic to suit your limits if they don't want to.

Immediately prior to departure, and recognizing the normal schedule of aviation weather distribution, obtain your final weather briefing in preparation for filing. You've roughed out a route that tells you which terminals to ask for in the forecast section, and you have a pretty good idea of your altitude. A talk with Flight Service can go very smoothly when you're familiar with the route; just throw a destination at the specialist and you've given very little cue to what sort of briefing you expect. They also need to know what sort of airplane you'll be flying, because the briefing you get will reflect its capabilities.

"Morning, Flight Service, Turbo Stationaire seven nine five five Hotel is looking at IFR Victor Six, Reno to Salt Lake City, leaving in about an hour from Reno Stead with a VFR departure, about three hours en route, and we'll need an outlook for a return tonight too." That gives them all the essentials:

Type (and implied capability)

Ident (really just tells them you're a pilot and fills their logbook)

VFR or IFR

Route

Destination

Departure point (you never know when a very local weather condition might make that element important)

Departure and en route time

Any other elements that might affect the briefing.

If you've listened to a TWEB or watched NBC's "Today" show weather map, tell the specialist. For example:

"I see you have a low aloft that may be affecting Victor Six up toward Salt Lake and the possibility of some frontal action later in the morning, according to the TWEB for the route. Can you fill me in?"

You've just saved one specialist some time by briefly mentioning your awareness of the general weather picture, and little things like that can make a big difference.

Now you're likely to get a detailed description of the most recent reported significant weather and the rate at which it is changing, both in quality and position. This will take the form of a narrative description, based on the current area forecast, something like this:

FSS: "Right, the low as last reported centered around an area just north of Victor Six at the Utah-Nevada border, and it's expected to move by this afternoon to a position generally southeast toward the Four Corners area."

Unless you take shorthand, you'll never get all that down. How can you make weather briefings easier to record and analyze? With a map, that's how.

Aviation writer Stephan Wilkinson once designed the perfect weather-briefing sheet, and I've photocopied the thing hundreds of times. It featured a map of the United States with the state borders and the location of most of the major air terminals. The same idea has been around in various forms, all of which recognize that a map is absolutely necessary to make sense of area weather briefings. (I keep two other versions of a briefing sheet, one from the FAA/GAMA/Ohio State University and another from Gibson Aviation Insurance in Gaithersburg, Maryland.) If you can't find a facsimile of such a map and briefing sheet, here's how to make one of your own:

Every telephone book that serves a good-size city contains an area code map of the United States. Take a piece of tracing paper and a felt tip pen and draw the outline, including the state borders. Take your finished map and find a photocopier service that will reduce it in size to about half or less of the original dimension (omit this step if you don't want to design a full briefing sheet but just a map). Paste the image of the reduced map on the top of an ordinary piece of typing paper and in the space remaining, design your own briefing sheet to include en route stations (present and forecast), destination and alternate weather (present and forecast), winds aloft forecasts, icing, Notams, and pilot reports. The boxes for icing and pilot reports can also serve to record cloud tops or any other conditions. A sheet organized by *you* instead of someone else will always make sense to you and will serve two purposes: a cue to ask the right questions during the briefing; a place to write it all down in logical fashion. After you've designed the sheet, have it photocopied and hang on to the original. You may want to redesign it later — to fit into your Jepp book, say — or have some more copies run off.

Carry several copies of the map and briefing sheet with you wherever you fly and use them to *draw* what the weather briefer describes. You'll find these maps particularly handy for depictions of weather radar reports of thunderstorm activity. Flight Service has to translate the picture of the storm activity into narrative, and they do it by geographic reference, often describing the four corners of a severe weather warning area. It's too much to keep straight in your head, but drawing it on an expendable map makes it easy to see exactly what's happening and get a picture of the dynamic weather situation.

Since I tend to fly some routes more often than others, I file the briefing sheets and maps along with the flight plan I wrote to go with them at the completion of a trip along a frequently traveled route. Over a period of time, you can decipher a pattern in the flying, and if you keep it up long enough, you can even draw a kind of seasonal picture of a route. The best single advantage to keeping old flight plans, though, is that you can just pull them out and copy down the route. If you find that the clearances issued for a particular route fall into a pattern, that's good information to hang onto as well. The point is, although these briefing sheets and flight plans are intended to be disposable, that doesn't mean you have to throw them away.

The rule book says you need file only one alternate, and there are even times when you don't have to file one. A quick way to save time and effort is this: *always* file an alternate. The rules don't say you have to use the alternate, you just have to file one that meets the minimums. The apparent intent is to force you to obtain a weather forecast for an alternate airport, because, simply stated, that's the only effect of it. Once you're in the air, ATC only wants to know what you intend to do now, not what you filed earlier. Therefore, on the matter of forecasts, you mentally operate on a couple of levels. There's the legal alternate, and then there's reality. Even the chance of low weather at the destination should be a signal to shift your thinking from mere legality to "What am I *really* gonna do?" In order to make that shift, you have to assess the weather as a system.

When you ask the weather briefer for forecasts at alternates, don't stop with a check of what you intend to file. It may be quite legal to file for an alternate airport that's five miles from your destination, and it may even be operationally attractive because it reduces your fuel reserve requirement to find an alternate close by. The trouble is, if the weather forecast for your destination turns out to be wrong, there is little reason to believe that an airport five miles away will be exempt from the mistake. Therefore, ask the briefer to give you some forecasts for terminals that are just plain *escapes*. Sure, they may be a hundred nautical miles off your track, but don't settle with Flight Service until you've picked one or two that are virtually guaranteed VFR or close to it. There is no situation where that won't work, even if you're flying along a really bad front and nothing looks good along the line of the route. Look to the sides, look anywhere, but *do* determine where your absolute total-disaster escape route will be. As they say about a certain traveler's check, don't leave home without it. It may turn out, surprisingly, that your real alternate will be a return to the original airport. For some reason, pilots are occasionally blind to the simple option of turning around in an airplane and going back to where they started; it just seems so humiliating.

Flight Service stations talk to the National Weather Service and to ATC through teletype codes, and although it's not important for you to know every code used in a flight plan, it helps them a great deal if you take the trouble to look up three-letter designators for airports that aren't well known. Jeppesen lists these in one remote corner near the upper left on the airport diagram behind any airport's first approach plate; they also appear in the J-Aid. The U. S. Government Printing Office also publishes a three-letter code book called *Location Identifiers*, but it's pretty expensive for that use alone. If the airport is private or off the beaten track, don't expect to get a forecast for it. They just don't list one for Birdwhistle Creek Flying Field and choice locales like it. You may as well look for the terminal nearest to such a destination because they're going to ask you anyway.

Always ask what type of airplane reported icing; most of the time they tell you, but bear in mind that an airliner may spend all of three seconds passing through conditions that could ground your two-place. Light rime coming from

144/PLANNING AND ORGANIZING A FLIGHT

a line captain may mean severe wing anchors in a Stinson. Same goes for turbulence reports; you'll find with experience that lightplane pilots tend to overstate the actual turbulence.

I always ask the man the current Greenwich time. First of all, it eliminates looking it up on one of those tables later, and it also serves to double-check the battery on my wristwatch. Flight Service can provide other helpful information when they're not terribly busy, as in the case of uncertain routing, where a radar service area lies, anything from the *AIM* parts to which you don't subscribe. They're supposed to tell you about navaids along your route, but back them up if they forget. I usually terminate any briefing by saying, "That's all I can think of. You have anything to add?" Some of the answers have been pretty surprising — in one case, a military exercise that I'd missed reading in my weekly chart revisions and which would have created a considerable delay at the very least.

Once you're satisfied that you have enough weather and route system information to allow you to plan a safe flight, you can divide that information into two categoris: flight plan information and flight *planning* information. You didn't think the planning stopped with these phone calls did you? Planning continues until you're tied down safely at the far end. But first, the flight plan.

ATC will receive the first eleven items on the flight plan, and the rest is kept by Flight Service. Thus, fuel on board, alternates, pilot's name, number aboard, and aircraft color never make it to the center and have no influence on your entry into the system; their main purpose is a bit grim: rescue in the event you don't show up as expected. For flight in controlled airspace at IFR altitudes and for flight in the Positive Control Area, an IFR flight plan is mandatory for entry.

Just a few pointers on the flight plan data blocks:

— aircraft type should include the manufacturer and the *model number;* they think it's just swell that your Grumman American is a "Tiger" but to the computer, it's an AA-5B, and your Cessna "Golden Eagle" is a 421, with A, B, or C suffix as appropriate.
— special equipment, use the correct code for *operational* equipment, so if your RNAV has closed its eyes, for Pete's sake don't file suffix C, F, or W (or whatever letters they're using for RNAV equipment these days); the equipment code affects the kind of clearance and handling you'll receive.
— true airspeed, not according to the brochure, but the actual calculated true airspeed for the altitude you'll file in block 7. Some twin-engine airplanes gain airspeed as they burn fuel; if you choose not to throttle back to compensate, average the airspeed for the midpoint of the flight.
— departure point and destination in three-letter codes (some are letter/number combinations, as Q49 for Firebaugh, California)

— departure time, in Greenwich Mean Time, and give them at least thirty minutes to process you if the flight plan has to go through the center computer; proposed, if you're on the ground; actual, if you're airborne

— altitude, your initial *en route* altitude, not some altitude that may apply for a portion of the SID or other departure procedure; where you plan to be when you announce straight and level; your computed airspeed should agree with this, by the way.

— route of flight

Let's stop right here.

You are trying to reach Flight Service, and there's only one station within range out here in the desert. Naturally, some idiot is air-filing a flight plan, which wouldn't be so bad, but here's how he gives the FSS his route: "Um, Victor Six, Lovelock, Battle Mountain, Dobys, Wells, Lucin, Ogden, Victor 21, Salt Lake City." At the Flight Service Station, the specialist started scribbling every one of those fixes down, then realized what was happening, balled the sheet of paper into the wastebasket, and sat back until the transmission had ceased. You had to wait while Flight Service instructed yet another pilot on how to file a route.

Airway routes are filed in segments defined by the Victor number, not by each navaid along the way. You therefore indicate the fix at which you enter the airway and the fix at which you leave it for another airway or a direct segment.

Every route of flight must incorporate some departure segment, and the ATC computers recognize the following:

— any published standard instrument departure (SID) if you file with its full name and any appropriate transition to selected airways, for example: Newhall Two Departure, Palmdale Transition, which, in code, looks like this to the FSS person: NUAL2·PMD.

— "direct" (insert any appropriate navaid or fix name); and intersections are allowed even though, strictly speaking, you can't actually navigate to those without area navigation equipment

— a radial from a VOR

— a magnetic bearing

— a localizer

Every route of flight filed (unless you're already airborne) must indicate how you're going to navigate from the takeoff to the en route or arrival segment. You do yourself a big favor by devising departure routes that incorporate the above-listed navigation modes, because there's a computer code for each of them, which means they can be accepted without a lot of hassle by ATC.

Some pilots get confused about the departure procedures that Jeppesen

146/PLANNING AND ORGANIZING A FLIGHT

publishes on the airport diagrams following the first listed approach plate. Because these procedures are grouped with the airline-transport minimums for departure, many pilots think these procedures are for the lines only; not so. It's true that these procedures apply to commercial operators, but they're for Part 91 flying too. It's just the minimums that apply to the commercial types.

So any route filed should include the following:

— a departure procedure to a navaid, fix, or airway that defines entry to the en route segment

— from there, the Victor number of the airway segment to be used, with names of fixes or navaids that define *each change* in airway, or

— if the flight will proceed direct between two navaids, the name of the navaid or fix at which the direct segment will begin and the navaid to which you'll be navigating direct

— a fix or navaid that defines the end of the en route segment and forms the transitional fix to an arrival procedure, with the actual arrival procedure *implied* by the choice of a fix, or

— a STAR, if you want to request one

If you prepare a route-of-flight entry that includes the information above, takes into consideration such variables as weather and such constants as preferred routes, your chances of hearing the wonderful words "clear as filed" go up logarithmically. Now make sure you take a copy of the flight plan with you to the airport in the event that the miracle happens! Back to the flight plan form:

— destination should include the name of the airport *and* the city (to avoid any confusion over which airport named "Oklahoma City" you really want) or the three-letter code

— estimated time en route should be the total time for climb, any departure procedure route, time en route at cruise airspeed (computed with wind correction for your estimated groundspeed) to your destination

— under remarks, add any notes that may affect the ultimate clearance, such as "VFR departure" or "climb VFR to (name of navaid or fix)" if you want to reduce delays.

So far as ATC is concerned, you've done your job. Add a legal alternate for the Flight Service Station and the rest of the data and you've honored your obligations to them. If you file a *composite* flight plan — one in which some of the flight is IFR, some VFR — nothing changes except that you tell them the point at which you'll make the change and check both VFR and IFR at the very beginning, under "type of plan."

When you've filed the flight plan, you can organize your bookkeeping for the flight. Various outlets sell all forms of flight logs with tables set up so you can list each fix and keep your estimates in order. You're not required to buy these, though, and there's a rudimentary log on the back of flight plan forms at the FSS. I usually use a clipboard and an ordinary pad of paper, lying flat and with the lines on the paper set up vertically so that they form columns. The advantage to setting up your own log is that you'll design it to *your* idea of how it should be organized, not somebody else's notion. With the store-bought logs, you end up glancing up to see which column is which all the time, and you'll write the estimates where the actuals are supposed to go and vice versa.

It also helps if you have set up a separate writing space for copying clearances before they read one to you. Organize this part of your office with a vertical column, like this:

> Cleared
> Route
> Altitude
> Frequency (or Departure)
> Squawk

The first element they'll read you will be the clearance limit. When things are going well, it'll be the destination airport — you have a reservation all the way — but there are times when they'll clear you short of that. Even if you've been cleared as filed, they read you the departure procedure anyway; you haven't done anything wrong, it's just part of their rules. If you are cleared for flight plan route, that does *not* necessarily mean anything about the *altitude* for which you filed, just the *route*.

You should already have a pretty good idea of what the departure frequency will be, and a lot of pilots just dial the squawk right into the transponder instead of writing it down. Almost anything handy in the cockpit can be used in a pinch to remember a number — tune the departure frequency into the radio, the altitude on the ADF card or the true airspeed adjustment. The real secret to getting a quick copy on a clearance is knowing a whole lot ahead of time, and since you can take a wild guess at most of it, you can probably fill in at least two and maybe three elements before the reading begins.

If a controller reads a clearance faster than you can copy it, have your own private shorthand symbol that means "missed everything after this and before here" and then just keep writing where you can pick it up. That way, you can call back and ask for just the missing element rather than the whole clearance over again.

Some advice on clearance-copying shorthand: a lot of people who are just learning instrument flying spend entirely too much time memorizing somebody else's idea of the "right" shorthand system.

There is *no government-approved shorthand system.*

Here are two rules that will help you copy clearances:

I. Design your own shorthand; you will remember it — it's *yours.*

II. Do not use somebody else's shorthand system; you will forget it — it's *theirs*.

There are really only a few key words you'll meet frequently that can benefit from some sort of quick method to get them on paper. One friend of mine just draws a wavy line for the words "flight plan route," and that struck me as being better than "FPR," which could, someday, get me confused with a facility code; the wavy line was easier, so I use that now.

You might like to invent symbols for the following:

radar vectors
runway heading
radial
bearing
cross
depart
intersection (make sure it's different from "cross")
intercept (make sure it's different from "intersection")
turn right
climb . . . and maintain
descend
turn left
hold
inbound/outbound
VOR/VORTAC
at or above (also, at or after)
at or below (also, at or before)
direct

Have a go at it. All's fair: letters, arrows, funny faces, whatever makes sense to *you*. Once you invent your shorthand, stick to it.

As you progress through the flight, you'll have to write the amended clearances somewhere, so make room for them. You may want to work to the right of your vertical column or simply write in the margin of your flight log. Never miss writing down a com frequency change, though; it's not at the change you'll need it but later, when the next change comes. If you write down every com frequency change at every handoff, you'll never have a problem if you can't raise the next station. Just look back in your notes, and even if you've tuned the last frequency out of the dial, you'll be saved. (Some pilots count on the newer radios' frequency storage capability to handle this chore for them. Just store the old frequency and tune the new one.)

Before you start the engine, get the cockpit organized for flying with a minimum of fuss. Your clipboard should have your flight plan, your flight log, and your weather information sheet within its grasp. Beneath the paper you write on should be the charts for departure and en route. Most experienced IFR pilots have the approach plates for the *departure* airport tucked in there with the SID so that the engine-failure-following-first-power-reduction of song and story can be followed by a perfect approach and landing. As you use charts, remove them from the clipboard and tuck them into a map pocket or back into the chart carrier. The point is to separate active from inactive charts.

A note about approach chart clips: some people like 'em, some don't. I would wager that it breaks down according to who subscribes to Jeppesen and who buys the NOS bound books. I tend to find the clips useful as *the* holder for all procedure charts. In other words, before the flight, they all get clipped up there on the yoke in order of use — SIDs, departure airport approach, destination plates — and they never get lost in the shuffle that way.

All airplanes are as different as people when it comes to finding a place for the IFR bookkeeping once you're under way. If you use a large chart case, you'll never find an airplane that offers space for keeping the whole thing handy until you advance to the cabin-class hardware; in a little airplane, pull out what you need and arrange it nearby. (I find charts on the glareshield very distracting in situations where I'm watching for traffic. The reflection of the white chart tends to obliterate the image outside.)

Never depart without a chart check. Almost everyone has faced a situation wherein the charts you need right *now* are in the nose baggage compartment or the wing locker. Luckily, you don't break any rules by calling ATC and admitting you've lost the correct chart, so don't panic if this happens.

I've never met an instrument pilot who denied that neatness counts — honestly. The longer you fly, the more compulsive you'll get about the housekeeping, and that's good. If you can accomplish an IFR chore early with no effect on the success of the flight, do it now rather than wait. If things go wrong, that little advance work can add up to a real face-saver. Even knowing where the sectionals are beneath the seat may allow you to reach down there without taking attention away from a no-gyro panel. What? Sectionals? You bet; they're part of your equipment, even IFR.

Every IFR flight that includes an en route segment ought to include at least one weather check with Flight Watch if your preflight forecast said you can expect to find instrument conditions on arrival. Flight Watch gets first look at any change to area forecasts that merits a Sigmet or Airmet, so if any have been issued, your chances of getting in on the information are best if you call them. It's also your last chance to establish any change in

— the general trend of the weather for better or worse
— the *rate* of the weather change, which is just as important as knowing the quality of the weather.

Part of planning well is allowing for changes in the plan. Just as you wouldn't adhere to an IFR flight plan if it were going to take you right through a thunderstorm, neither would you push ahead to an approach if it were obvious that your destination will be out of business. I want to know as much as I can as early as I can, and that includes planning considerations for the possibility of canceling the original destination before "having a look."

If you're working with ATC under IFR, whether you're on top and cleared VFR or not, you can talk to them about basic changes in your flight plan without going back to Flight Service. Mind you, in order to leave the ATC frequency to check on weather (usually with Flight Watch on 122.0, which operates solely to provide weather update advisories for routes and destinations), you have to check in with ATC first. Suggest that you'll keep the time you'll be off frequency to something like five minutes maximum, and they'll usually ask you to give them a call once you switch back onto the ATC frequency. If destination weather makes it appear that an approach would not be advisable, you needn't call Flight Service to amend the flight plan. Discuss it (after you have a clear idea of your alternative plan, of course) directly with the controller. Since the major consideration at this point will be to fit your new route into existing traffic — and chances are that in bad weather *everyone* will be trying to get a change — the ATC controller is the only source of help. A controller is also the best source of real-time weather reports, such as the required pilot report from the last aircraft that was unable to complete the approach and then executed a missed approach procedure, or that had to divert from a feeder route for a local thunderstorm.

Changing your plan is easier if you planned well in the first place. Your study of the situation the night before really pays off when an amended clearance comes through with one weird-sounding intersection name after another. You've already scanned the en route and area charts and read the names at least once and traced the major airways in the neighborhood, making a mental note of their Victor numbers. Your confidence in your airplane's rate of fuel burn, the result of record-keeping over a period of time, pays off in a firm estimate of your endurance and range, given the actual winds you've encountered, then computed, based on course and heading information.

Your prior knowledge of "speed factors" — the poor man's DME — have enabled you to come up with rapid estimates of time to various destinations without distracting yourself too long from the job of attitude flying, especially when the weather is getting worse. When the moment calls for it, you've asked ATC to confirm your position so that you can increase the accuracy of those estimates. At moments like these, you will come to realize that IFR flying is really one long plan. Even your flight log accomplishes more than just providing estimate data. From it, you gain your real assessment of how the wind is affecting your endurance-range equation. You gain an accurate picture of how *time* (which is all the fuel tanks give you, really) translates into *dis-*

tance. Without even thinking much about it, that knowledge becomes a plan. It either confirms your original destination or it revises it. I would suspect, without knowing, that many IFR pilots get themselves into bad situations with weather they can't handle because they think the planning stopped back on the telephone. They think a flight plan is inviolate, engraved in zinc, and placed in a vault at the Treasury Building. If you can be persuaded that a flight plan is really a "reservation" for airspace, that the planning really continues throughout a flight, that the process of estimating never stops, you'll begin to understand the reason behind all the bookkeeping and the hours spent the night before a departure. If you can help it, start from the beginning to plan every IFR flight as if you were carrying paying passengers. Don't just please your instructor or pass an exam or a flight check.

I remember being taught flight planning for IFR at one of the larger flying schools, and it was only after I'd gotten clear of the place that I learned how experienced IFR pilots really plan: they estimate.

Now don't get me wrong. If your instructor is the type who wants a multicolumnar estimate sheet, complete with true airspeed, magnetic variation at each navigation fix, wind triangle recomputed every ten minutes based on the forecast converted to true and divided by the deviation, there's nothing to do but fill the thing in. Ask yourself this question, though: if you were in an airplane in an absolutely wretched waterfall of a thunderstorm and you had to revise that plan, would you do it the same way? Of course not.

I wish I'd been taught from the first to estimate well. Oh, you find that you manage to get better at it as you go along, but more than anything, I'd have valued an instructor who'd have taught me the tricks of speed factors and demonstrated to me how accurate an estimate can really be. I'd have had much more confidence in it, and it would have spared me proving to myself that over the years, if you convert the wind at altitude into Kentucky windage and add or subtract from your usual average groundspeed to account for wind, then divide that estimated groundspeed into the distance, add a few minutes for climb, you're going to be within five minutes of your actual touchdown 90 percent of the time. The other 10 percent of the time, the wind forecasts are all off. It may be the best-kept secret in the United States, but almost the entire active IFR flying population gives ATC its en route time in exactly that fashion.

If you want to be doctrinaire about it, go ahead: call it sloppy, accuse the radar environment of allowing pilots to slough off on planning. But if you step back and really look at what's happening, you'll find something below the surface that's startling. Pilots who plan from estimates are in better shape to fly IFR when something unexpected happens. They have a dynamic sense of the airplane and its real performance within the surrounding airspace, whereas the pilot who fumbles with numbers to satisfy some training regimen that taught accounting instead of estimating is lost as soon as he or she is deprived

of the time for the chores. You can't simply figure some super-accurate estimate to the next navigation fix; you have to be able to extend every situation to
the *destination*. That's your real job in the IFR front office, but you'll find no
flight-log planning sheet that really conveys that simple truth. All your
planning, no matter when it takes place, focuses on landing the airplane, and if
you do not plan *all the time* with your ultimate solution the *whole trip*, you
may find yourself someday arriving precisely at some navaid that's 90 NM
from an airport at exactly what your estimate predicted, and with compass
aligned at the precise number you wrote in the little box in column four — and
with 65 NM worth of fuel in the tanks.

Let's take yet another look at that flight director's asymptotic curve and try
to understand it in terms of rate of change from a somewhat different point of
view:

If a flight director could talk, it might say things like this as it arrives at the
solution to its course-correction problem.

"Seems like the course line is over here to the left, so let's crank in some left
bank and see what happens.

"Yup, uh huh, things are looking better. Bank and heading change appear
to cancel out the course displacement. No question. The course line is getting
closer, so let's look at *how fast* we're closing on it. Whoops, better ease off a
little bank. There. That slowed the closing rate. Almost there now. Close
enough. Resume straight and level."

Learning to plan by estimating is like that. It's like a series of large-scale so-
lutions that are progressively refined, getting closer and closer to the ideal,
until it finally arrives at the answer that's so close to the actual that the difference can no longer be measured. The asymptote, you see, *never actually
touches the course line*. But it comes so close, and that's what matters.

IFR Notebook

The Gospel according to FAR 91.5 sayeth:

"Each pilot in command shall, before beginning a flight, familiarize himself with all available information concerning that flight."

In other words, if it's "available," you're supposed to be "familiar" with it.

Minimum planning elements for IFR flight:
  weather
  fuel
  alternates
  known traffic delays
  runway length
  landing and takeoff performance

Fuel Rule

You may not *operate* in *IFR conditions* without enough fuel to

complete the flight to your planned destination fly to the *filed* alternate at any airspeed you choose fly after that for 45 minutes at normal cruising speed

If

the destination has an instrument approach (published) *and*

the forecast for the destination airport covering the period from one hour
  before to one hour after your ETA indicates a ceiling at least 2,000 feet
  above airport elevation and visibility of three miles (it doesn't say
  whether they mean nautical or statute but statute is lower), then

You may fly with sufficient fuel to
   complete the flight to the destination airport
   fly after that for 45 minutes at normal cruising speed.

Flight Plan Rule

You must file an alternate airport unless

the destination airport has a published instrument approach
and
weather reports or forecasts in any combination indicate from one hour before to one hour after your ETA a ceiling at least 2,000 feet and visibility of
   three miles.

Alternate Minimums Rule

In order to qualify as an alternate airport in an IFR flight plan, an airport must
meet the following conditions —

   if the airport has a *published instrument approach*
      current forecasts indicate that on the *ETA* at the *alternate airport* (not
         the destination), the ceiling and visibility will be at least the published
         minimums for an alternate airport as listed on the airport side of the
         first approach plate or in the minimums section
   or, if no minimum is listed
      for a precision approach (ILS/PAR), at least
         ceiling 600 feet, visibility 2 statute miles
      for a nonprecision approach (VOR/NDB), at least
         ceiling 800 feet, visibility 2 statute miles

   if the airport has *no published instrument approach*
      current forecasts indicate that upon ETA at the *alternate airport*, the
         ceiling and visibility will be at least
      VFR conditions sufficient to allow descent from the published MEA,
         approach, and landing.

Warning: VFR conditions require at least 500 feet distance *below* cloud;
therefore, a descent from MEA in VFR conditions would require that the
ceiling be at least 500 feet above the MEA.

Weather Report Schedule for the U.S.

airborne	call Flight Service
ground	call National Weather Service or Flight Service

Sequence Reports
surface weather observations made hourly and available
at :10 past the hour
VALID: one hour

Terminal Forecasts (FT code)
a 24-hour forecast, with categorical descriptions for the last *six* hours'
outlook as:

(LIFR)	low IFR: ceiling less than 500 feet/visibility less than one mile
(IFR)	IFR: ceiling 500 to 1,000 feet/visibility one to three miles
(MVFR)	marginal VFR: ceiling 1,000 to 3,000 feet/visibility three to five miles
(VFR)	VFR: ceiling more than 3,000 feet/visibility greater than five miles

East of the Rockies: 0940Z, 1440Z, 2140Z
West of the Rockies: 0940Z, 1540Z, 2240Z
VALID: to 24 hours

Route Forecasts (for Pilots Automatic Telephone Weather Answering Service [PATWAS] and for Transcribed Weather Broadcasts [TWEB] as well as weather briefings by FSS)
forecasts issued three times a day for TWEB routes between principal
terminals
VALID: morning and midday   12 hours
               evening   18 hours

Area Forecasts
an 18-hour forecast, with a categorical (LIFR, MVFR, etc.) outlook for
the 12 hours following, organized by regional descriptors, such as names
of states, not by terminal designators
Issued every 12 hours at 0040Z and 1240Z
VALID: 18 hours

Winds/Temperatures Aloft Forecasts
12-hour forecasts of wind direction and air temperature at 3,000-foot altitude intervals to 12,000 feet (except no temperature for 3,000 feet), then
at 6,000-foot intervals to 30,000 feet (and it goes higher) for about 120
locations
data prepared twice daily, at 12-hour intervals and
VALID: nine hours

Sigmets:
>   weather advisories describing conditions that are hazardous to all aircraft
>   categories, including
>>   thunderstorms, tornados, hail
>>   severe or extreme turbulence
>>   severe icing
>>   duststorms or sandstorms lowering visibility within 150 NM of FSS
>>   broadcast at 15-minute intervals at :15, :30, :45, and :00
>   for weather briefings, as received
>   ATC facilities: alert notice only broadcast to all aircraft

Airmets
>   weather conditions hazardous to single-engine and light aircraft,
>   including
>>   moderate ice or turbulence
>>   winds of 30 knots or more within 2,000 feet AGL
>>   visibility of less than 3 miles, including mountain passes and ridges
>>   obscured
>   also, hazardous conditions to *all* aircraft, including
>>   continuous low ceilings/visibilities
>>   moderate turbulence over mountains within 150 NM of FSS
>>   broadcast at 30-minute intervals   :15 and :45

Alert Notices
>   at 30-minute intervals   :15 and :45
>   refer to Sigmets or Airmets

Weather Radar Reports
>   scheduled observation at :40 or more often, whenever precipitation is
>   being detected
>   plus
>   hourly radar summary in plain language
>   plus
>   16 facsimile charts per day at FSS (in person briefing)

Weather radar is considered to produce a very reliable depiction of areas of
precipitation within 100 miles and of intense weather at distances greater
than 100 miles
>   weather radar echo intensity levels: 1 through 6

Transcribed Weather Broadcasts
>   continuous tape recorded, changed irregularly
>   facilities marked "TWEB" on charts — NDBs and VORs
>   locations within 400-mile radius of broadcast controlling station
>   general weather, Pireps, radar, winds aloft

Scheduled Weather Broadcasts
:15, by Flight Service personnel
weather reports, Notams
NDBs and VORs
within 150 miles from broadcast station

Flight Watch
single frequency 122.0 MHz
Flight Service personnel
6 a.m. to 10 p.m.
5,000 feet AGL
primary Pirep input inflight

Route Structure Planning

brief format, see text for explanation
DEPARTURE (Required)
SID*
direct* to fix or navaid
radial*
magnetic bearing*
localizer*
departure procedure
VFR
radar vectors
dead reckoning
EN ROUTE (May be omitted on short flight)
airway* (Victor, J or other)
direct*
radial*
magnetic bearing
radar vectors
VFR
dead reckoning
ARRIVAL (Required)
STAR*
"implied"* (by the final airway route and fix)
contact/visual
VFR
radar vectors

computer compatible — most easily processed by FSS into a flight plan

Speed Factors, or Poor Man's DME
To obtain the time to fly a certain distance,

If your groundspeed is	multiply mileage by
60	1.0
65	.9
75	.8
85	.7
100	.6
120	.5
150	.4
200	.3

CHAPTER VIII

Procedures

Within the context of IFR flying, the word "procedures" connotes "complexity." Approaches to airports are "procedures," as are certain choreographed departures that follow a published routine. Pilots prepare themselves mentally for procedures as if there were something wholly different about them that separated their range of difficulty from the rest of IFR. Actually, IFR itself is just one big Procedure with a capital P; it might just as well be called IFP, because the entire regimen of instrument flying and navigation is simply a disciplined way of doing things according to published rules.

"Procedures" also connotes charts, as the bulk of the printed material you deal with as an IFR pilot involves approach plates. The correct use of charts and the knowledge of the range of information they contain is a study in itself. James Terpstra, of Jeppesen, has taught a course entitled "Charts and the Airspace" to S.R.O. audiences all across the United States, and yet there are still pilots who may not have a thorough knowledge of the Jepps. Government publications (Advisory Circular 90-1) instruct pilots in the use of NOS-published approach charts, yet there are hidden facts that charts do not reveal but which pilots really ought to know. The Department of Transportation promulgates the *United States Standard for Terminal Instrument Procedures*, otherwise known as the TERPs, and this inexpensive beige-covered book is the *real* bible that describes the structure of various procedures and the rationale behind them.

We use certain understood and agreed-upon "procedures" whenever we encounter the ATC system, rules like required reports and mandatory two-way communication that spin off entire worlds of rules and recommendations ("Recommendations" have all the impact of a rule when something goes wrong and the insurance company gets involved). Up to the point of proximity to the airport, though, the word "procedures" has not really connoted the integration of all elements of IFR flying into a single task that calls upon the ability to fly an attitude on instruments, navigate to a new set of performance standards associated with *approach* category navaids and an increasingly refined standard for aircraft positioning.

Procedures are like funnels. Even departure procedures are like funnels, but upside down; they enable you to exit safely from an area of terrain and obstacles to the relatively clear controlled airspace of the en route system. Arrival and approach procedures funnel you downward in a descent from airspace with relatively generous assurances of obstacle clearance to areas with stingier guarantees. Procedures can therefore be thought of as that phase of flight wherein altitude and its associated pitch and power control are much more the predominant axis of precision. That's not to say that directional control becomes any less important but that any *error* in altitude control has far graver consequences and associated risk.

Some pilots regard the task of maintaining precise altitude and terrain clearance with such awe that they think the proper solution is to *add* altitude to the published minimums for various segments of any procedure — approach or departure — and thereby assure themselves of an extra buffer. The trouble is, ATC generally assumes that any pilot will fly at the published minimums if a controller doesn't specify otherwise. Therefore, on departure procedures with altitudes specified, those altitudes are tantamount to clearance instructions, and changing altitude on your own may cause you to intrude on another airplane's airspace. Similarly, adding altitudes to the published minimums on descent and arrival procedures may make parts of the procedure more difficult to fly. It is generally assumed — and this makes procedures somewhat different from en route navigation — that you will begin descents on your own at each appropriate fix and that you will continue the descent down to the published minimum as expeditiously as possible, then maintain the minimum. Once established on an approach, ATC will provide you with no further altitude clearances until you should report missing the approach. Functionally speaking, the difference in procedures is that you will probably be changing altitude more often, that you will fly altitudes *exactly*, and you will therefore refer more frequently to the altimeter, rate-of-climb indicator, and airspeed indicator — the instrument group that gives you all your most accurate pitch and power information.

The structure of procedures, as published in the TERPs, takes into consideration two factors: the situation of terrain and obstacles near the procedure airspace, and airplane performance. (Even departure procedures for non-Part 91 operations consider the airplane's performance.) By describing minimum standards for clearance from obstruction, the TERPs essentially tells an airport whether an approach can be installed there or not. The approach, in other words, is fitted to the real layout of the airport, and if some obstacle appears within the airspace specified to be clear by the TERPs, either another type of approach with more generous clearances must be used (usually with higher minimums) or an approach may not be feasible at all.

Pilots generally are informed only of course, navaid, and altitude information. Neither Jeppesen nor NOS charts provide structural information about procedures, and for a very practical reason: it would clutter any chart to the point of illegibility. Once you examine the criteria listed in the TERPs, you begin to understand the magnitude of the problem of assuring obstacle-free airspace. None of this is meant to frighten you. The scale of the approach structure must be clear to you, though, so that you understand that at the outer marker, a two-dot fly-up deflection on the glide slope puts you about 100 feet low in an area with, at best, 500 feet of obstruction clearance and perhaps as little as 200 feet! We've provided some structural information about approach procedures, including missed approach segments, herein; to understand fully how approach airspace is defined, however, you should obtain and read a copy of the TERPs.

All approach procedures are designed to vary in dimension with the category of airplane, and there are five categories, A through E, defined by weight and an airspeed factor found by multiplying the stall speed in the landing configuration at maximum gross landing weight by 1.3; an airplane that exceeds *either* the weight or airspeed criteria must apply the *highest* appropriate category that matches the speed or weight standard.

Here's how it works: your airplane weighs less than 30,001 pounds, which would put it in Category A, but you may want to fly it faster than 91 knots indicated during, say, a circling maneuver, and 91 knots is the maximum allowable for the 1.3 Vso figure in that category. Even though the airplane's weight would put it in Category A, its actual *computed* operational category is determined by your indicated airspeed during the real approach. For circling, you'll probably feel a lot more comfortable flying this particular airplane at 100 knots.

You've got yourself a Category B airplane there, bub, and you *must* apply the somewhat higher minimums that may prevail, particularly for circle-to-land maneuvers and when certain components of a precision approach system are out of service. Now you know why some makers of high-performance business jets are so anxious to lower the airplane's stall speed. You never go around stalling in jets; it's just to get you into more airports in low weather.

In general, approach and departure procedures rely on positive course guidance. The exception may be the initial approach segment, in which case the obstacle clearance area is simply expanded; a more precise way to express it is to say that you'll have to look for a place with a bigger open area, because that's what it boils down to.

Generally, departure procedures are constructed in such a way as to avoid adjoining terminal airspace or nearby terrain and to provide an orderly and simple means whereby a pilot can find his or her way from a visual takeoff to the airway or en route structure. The procedures therefore operate as standardized clearances, self-contained little navigational cha-chas that allow you to take off using a particular runway, fly fairly directly to some form of course guidance, which may be radar vectors for a portion of the procedure until you are established on the centerline of an airway or radial and with sufficient altitude to derive reliable nav signals. Around mountainous terrain, designing such a SID can be a tough chore.

SIDs are usually constructed with a fundamental procedure that applies from takeoff to some fix or navaid. Takeoff procedures are devised to apply to each active runway. This is followed by a "departure" segment which may be followed by a "transition." When you are planning a flight (remember, plan backward), you determine which airway you wish to join en route, find the transition among the SID charts that leads to a navaid along that airway, then read the departure and takeoff procedures that precede the transition. The takeoff and departure segments may have a single name; individual transitions are also given names, usually to match the navaid they use. SIDs also may specify minimum climb performance, usually for terrain clearance within a certain segment. The minimum climb requirement appears in fine print, and unless you are aware of it and alert to check *every* SID for the climb spec, you may find yourself getting a call from a concerned controller — or worse, *not* getting a call. For your first act of chart preparation, take a yellow felt marker and thoroughly examine your SID publication — whether Jeppesen's charts or NOS's books — and accent any procedures that even hint at some climb difficulty for your airplane. Compare altitudes with rates of climb, and the answer should be obvious to you. If you're unable to comply with the specified climb in a SID, say so in your flight plan under "remarks." Some airports have simpler departure procedures that don't require a chart or diagram, and you'll find these published on the airport chart in the lower left-hand corner on the Jepps, and listed by airport in the NOS *Instrument Approach Procedure* books. Remember, though, that ATC doesn't have these procedures in the computer, so you have to file them by describing the procedure (direct to facility navaid, then left turn to 020 degrees magnetic, etc.) rather than labeling the whole mess "IFR procedure as published." It may be "published" to you, but it's not to them.

If a SID includes a crossing restriction requiring that you climb to some minimum altitude prior to reaching the appropriate crossing fix, ATC assumes that you're able to assure a sufficient rate of climb to the crossing fix *and that you'll continue to climb at a minimum of 152 feet per mile thereafter.* The structure of departures is not as thoroughly detailed as that for approaches, but in general, departures are very similar to missed approach criteria — which is exactly why missed approach procedures make such a nifty impromptu departure procedure if, by some chance, none is published for the airport. The so-called departure surface — a theoretical three-dimensional planar gradient that defines the obstacle-clear airspace — has a slope of 40:1. There are approximately 6,070 feet to a nautical mile, rounded off a little, and the surface slopes upward a foot for every 40 feet of horizontal travel — or 152 feet at the end of a nautical mile. That's where the climb minimum comes from. The rule has not enjoyed the privilege of any awesome publicizing. Estimate your airplane's ability to achieve that gradient by applying the groundspeed factor (see IFR Notebook) to 152.

You may think of departure procedures as being easier flying than an approach, say, because you're climbing *away* from restrictive airspace and into expanding maneuvering room rather than the more confining airspace, as you do during approaches. If departures are so much easier to fly, why do so many pilots fly them so poorly? Perhaps part of it is the mental letdown associated with an "easy" procedure; I don't really know. I suspect a large part of it is caused by taking off before being really ready. There's nothing worse than starting a flight already behind the airplane, yet it happens to people with alarming frequency. One instructor I knew used to drive home the point with a physical reminder to his student. As soon as the airplane was cleared to taxi into takeoff position, this guy required the student to "fly," that is, to have one hand on the yoke and the other on the throttle. No more paperwork or fussing with mike wires allowed; if you forgot something, you had to taxi clear and beg another fresh takeoff clearance. You weren't ready to fly, in his book.

When you hear the words "clear for takeoff," you should be able to recite your initial heading after takeoff — runway, right turn to _____, left turn to _____ (those are your *only* choices), and your initial altitude clearance. The initial navaid must be tuned and with the appropriate course selected, something that should be done during the pretakeoff cockpit check. Only two procedural matters should intrude as the wheels roll down the white stripe: transponder to proper reply mode and recording of takeoff time. If you have a copilot, even if it's just a passenger with no ratings, put the right-seater to work writing down the takeoff time. If you can't manage that, say the time aloud to yourself and *remember* it; this is no time for paperwork. In these first few moments of flying, you will be making an abrupt mental transition from the ground-bound to the airborne, you'll be experiencing some acceleration sensations that may disorient you, and the requirements of your first moments of attitude flying absolutely demand that you have the first portion of the departure procedure memorized.

Rare is the opportunity to fly through an entire published departure procedure, because as soon as you contact the appropriate radar departure facility, the odds are that the controller will vector you according to local traffic flow and other conditions. Still, there is almost no excuse for not at least getting yourself generally headed in the correct direction after takeoff. Failing to do that is an absolutely dismal way to start a relationship.

Although the structure and obstacle clearance criteria for departure procedures remain vague, it's implicit that separation from traffic and obstacles is most critical during the first few moments of a departure. If there's a turn following takeoff, you may reasonably conclude that the turn is there to direct you away from nearby terrain or the airspace of another airport. Examine all the procedures for a particular airport and you can decipher a pattern in the flow that tells you most of what you need to know about where, at all costs, you should *not* direct your airplane.

In a situation wherein you are called upon to design your own departure procedure, first look for avenues of visual exit. Consult the sectional charts and determine a course and distance that will put you 1,000 feet above obstacles within four nautical miles — that's airway clearance, and it ought to be adequate; in the mountains, make it 2,000 feet and you'll have to work out a profile-view gradient as well if you want to ensure that you'll be proceeding upward at a sufficient rate. You may draw your procedure on a separate piece of paper or mark it on the sectional itself, with time checkpoints (one minute, two minutes, etc.) for your dead-reckoning distance fixes clearly marked along the course line. The sooner you're able to make radio and radar contact with ATC, the better. Continue your monitoring of position, though, because you may be more aware of the exact terrain configuration than the controller is.

When you receive your clearance prior to departure, the controller is supposed to state the name of the applicable SID even if you filed it and the clearance is to be "as filed." (That doesn't apply to STARs, by the way; they're considered part of the route filed — they're even called Standard Terminal Arrival *Routes* rather than "procedures," and that's the way ATC looks at them.) If you are assigned a heading to fly at takeoff when you get your clearance, the purpose of the heading may not be stated if you're to be operating in a radar environment. In this situation, it is implied that the heading will be a prelude to vectors and is intended to flow you into a vector system currently in operation.

Some SIDs specify lost-communications procedures; when none is published with a particular SID, the standard lost-communications procedures apply as for an en route segment.

STARs and Profile Descents are specific routings that conform to ATC's requirements for traffic flow, usually around fairly busy terminals. STARs are merely intended to make clearance delivery's job a little easier by cutting down on routing chatter. Profile descents are designed for high-performance jet traffic, as a transition from the J-route structure to the terminal environ-

ment, and there's little chance your Cherokee will be expected to cross
Bryson at or below FL230 and at or above FL190 at a speed of 250 knots.
Lightplanes can expect to receive STARs in a clearance, though. STARs are
filed using a code that's very similar to the SID system except that the *transition* code precedes the arrival code, the reverse of the case for a SID. The
codes are right on the graphic chart, whether you use NOS or Jeppesen. So
the Barrett Lake One Arrival Imperial Transition into San Diego, for example,
codes like this:

IPL.4JL1

Instrument approach procedures have, by comparison with departures, a
structure that's defined in infinite, mind-boggling detail in the TERPs. And it
turns out that the TERPs-eye view of approach procedures helps to clarify the
way you fly them.

An instrument approach procedure may have up to four segments, each
with an individual function:

— initial
— intermediate
— final
— missed approach

There's also a circling approach area, but that's not considered an approach
segment. When an airport sets about constructing an approach, it must define
the final approach first, then build an intermediate and initial segment onto it.
We'll take them as we fly them, though.

The initial approach segment begins at the initial approach fix, which may
not appear on the approach plate at all, or only as a tiny "IAF" near the name
and frequency of the navaid that defines the fix. Not every approach has an initial segment. The function of the segment is to provide a route along which the
aircraft can transition from the en route phase to the intermediate segment.
An initial approach may be made along an arc, radial, course, heading, or radar vectors — or any combination of these. No initial segment can require a
turn onto the intermediate segment of more than 120 degrees, and if the turn
is to be more than 90 degrees the procedure is supposed to specify a lead radial
so that you can initiate your turn onto the intermediate segment with sufficient warning. An initial approach segment may be a procedure turn or a holding pattern descent, as well as a straight or arced course.

The intermediate segment is intended to provide an opportunity for the pilot to configure the aircraft and adjust speed and position for entry into the
final approach segment. Basically, it's expected that the airplane will be generally aligned and settled down during the intermediate segment, with the
gear down and visually checked and all distracting prelanding chores dis-

pensed with, so that on entry into the final approach segment, no turn greater than 30 degrees ought to be asked of the pilot, and the final approach descent can be generated merely by reducing the power setting to approach descent. Final approach flying operates in a navigational geometry in which rates of change go up quite rapidly — it's a "high-gain" environment. Therefore, nothing should distract you from the basic flying and instrument monitoring chores.

There is a school of instruction that teaches pilots to generate the final approach descent at the termination of the intermediate segment by lowering the landing gear in a retractable airplane, and there's nothing wrong with that except that it's a lousy time to deal with the yaw and pitch changes incurred by gear lowering just as you're trying to fly precisely. Dropping the gear also requires a visual check of down-and-lock lights as well as other visual indicators, and it's been my experience that even so brief a job as that can start a final approach procedure off with me behind the airplane. Entering final with the airplane just the way you want it to be for its trip down has a lot to be said for it.

The intermediate segment begins at the intermediate fix, which is not labeled except on Canadian Jepps, and ends at the final approach fix, which both Jeppesen and the NOS symbolize with a Maltese cross on the airport profile view. Because of the intermediate segment's role in providing a space for speed adjustments, the TERPs specifies that the descent gradient within this segment should be as flat as possible, with an optimum of 150 feet per mile and no more than 300 feet per mile. Obviously, the faster you've allowed your descent speed to build up, the worse you're going to have it on making such adjustments. As frequently happens, a controller may request something like "keep your speed up" even when you're well into the intermediate segment and would prefer to be making your adjustment for final approach airspeed. Either you have faster traffic following you or he's trying to fit you into flow from other directions with ample airspace separation. Some pilots interpret the request to "keep speed up" during an approach segment as a mandate to add airspeed. In actuality, the controller is merely asking that you not *slow any further*, and if you increase airspeed, you may screw things up just as badly as you would if you were to slow down. Try to comply with the controller's request, but *do not* exceed safe limits to do so. No ATC personnel are empowered to supplement an airplane's structural strength while it's in flight so that you can plummet through low-level turbulence beyond the yellow-arc airspeed. Speed control requests impose directly upon your responsibility to fly safely, and there should be no question in your mind as to your ultimate authority. All you have to do if you are unable to comply is to tell the people on the ground. There are other ways for them to make adjustments.

Procedure turns are one form of initial approach segment, and they accomplish two things, one of which is generally known and the other less well publicized: they provide an orderly means of course reversal; they offer an excellent opportunity to reduce airspeed by adding to the drag factor in the turn. Procedure turns are confusing, however, because the way they're expressed on approach charts may not define how they're supposed to be flown. The opportunity to fly an entire procedure turn is rare now that radar vectoring to approaches has become the norm in the United States. Even in the rare case of an approach in a nonradar environment, any feeder route or initial approach segment that leads an aircraft generally into the final approach course with only moderate heading change and altitude loss for the intermediate segment transition will specify "NoPT," in which case, you're not allowed to fly one even if you want to (unless you specifically ask).

One form of procedural course reversal is accomplished by flying *outbound* on the reciprocal of the final approach course and at a higher altitude. Upon passing the final approach fix *outbound*, you note the time and *as soon as practicable*, but generally within two minutes, initiate the course reversal by turning to the published *procedure turn side*. About 45 degrees is considered the normal heading change. There is no rule that stipulates exactly how such a course reversal must be made, however; there are pilots who fly a 90- to 270-degree course reversal, though I can't really understand why unless it's just to be different. The point is that the procedure turn side of the course has been set aside as clear of obstacles at the course reversal altitude. It's there to allow you to make any form of course reversal you like.

On occasion, you'll encounter a course reversal with either a teardrop or a holding pattern depicted instead of the 45-degree heading change. If either of those two patterns appears on the chart, you *must* fly the course reversal as indicated; both patterns imply tighter obstacle clearance restrictions. In the case of a holding pattern course reversal, you initiate an immediate 30-degree teardrop entry from the outbound course and pretend you're entering a hold. A parallel entry is also allowable, but I think it's more work.

Outbound legs after the first heading change should be between 40 seconds and one minute, with adjustment for the wind (residents of New Mexico take note — for you, with a headwind, it may take 20 minutes!). In no case should your estimated distance outbound exceed the mileage limit published on the profile view of the procedure turn. Ten nautical miles is normal.

The minimum altitude at which you should perform the course reversal is published in the profile view of the approach procedure, and although it may not be terribly obvious from the diagram, you are not supposed to descend to a lower inbound course altitude until you are established on that course with the wings level. Although some procedure turns are specifically designed to allow you to shuck a huge chunk of altitude, the descending is supposed to be done on the outbound and inbound courses, not during the reversal maneuver. That rule applies for course reversals and holds in situations other than approaches too. (Both procedures have roles during departures and en route segments as well.)

The sequence goes like this: cross FAF outbound and note time, initiate course reversal heading change, and fly outbound for about a minute, initiate standard rate 180-degree turn to reciprocal heading of that used for outbound course reversal. When the CDI needle comes alive, intercept inbound course and heading, adjust for wind correction, and initiate descent if one is required.

If the procedure turn is placed close to the FAF, you may have to hurry to get down to the next altitude step after you've reintercepted the inbound course. Use the turn to load the wings and bleed off as much airspeed as possible and practical so that you're as slow as you can safely fly for the ensuing descent. If there's a single element of arrival procedures that will give you more of a problem than any other, it's rapid descents in intermediate segments. I know, they're not supposed to be steep, but tell that to the chaplain when you're on the VOR approach to Santa Monica Airport and they've held you a little high at Geste Intersection. Ah, but no more war stories. . . .

Exiting from a procedure turn is usually a one-two process: get back on course and lined up for final; get *down* quickly so that you can settle in at the intermediate altitude and adjust airspeed prior to the FAF. Airplanes descend quite steeply when they're flown slowly, so the technique here is to resist the temptation to push the nose down — that only *raises* the airspeed and ruins your steep gradient. All's fair; anything you can find to add to your airplane's drag in that situation will help you to achieve a steeper gradient when you have to.

Airspeed in procedure turns is limited to 250 knots indicated, which ought to present no problems until you graduate to jets. Where a holding pattern is depicted for the procedure turn, remember to adhere strictly to holding pattern procedure; it's implicit in the depiction of the holding pattern rather than a standard procedure turn that a one-minute outbound leg becomes the outbound distance limit rather than the 10 NM that characterize the standard procedure turn. Even if you were flying at just under 250 knots in a holding pattern procedure turn, that one-minute outbound leg yields just over 4 NM; that's even less than the 5 NM leg allowed for Category A-only operations.

It is assumed by ATC that you'll make just one circuit within a holding pattern course reversal and then continue inbound. You may use the holding pattern for an additional circuit if you want to lose altitude down to the published procedure turn minimum or to get yourself settled down better for final. All they ask is that you *tell* them at the time you get your approach clearance.

Certain teardrop course reversals also fail to clarify exactly where the intermediate segment begins. In such cases, you are to assume that the intermediate segment begins at the point where you complete the course reversal turn and level your wings inbound, if the navaid for the approach is on the *airport*. So where is the final approach segment? It's considered to begin at 10 NM from the *navaid*. Similarly, for a navaid located off the airport that serves as the FAF, the *intermediate* segment is assumed to begin 10 NM from the navaid/FAF, with the final approach segment defined, naturally, by the fix.

On extremely rare occasions, an airport may conduct timed approaches from holding patterns at a final approach fix. The conditions that must be met in order to conduct such approaches read like the fine print in a lease, so don't expect to encounter this procedure too often. Here's the basic problem: ATC will assign you a time to depart the fix inbound. It's your job to adjust your circuits of the holding pattern to ensure that you're inbound over the fix at the assigned time. It's really no big deal if you break it down this way:

One holding pattern equals:

one turn at the fix end	1 minute standard rate
one holding side leg	1 minute
one outbound end turn	1 minute standard rate
one nonholding leg	1 minute

That's in no wind, of course. In general, you fly another holding pattern any time your clock time at the fix shows that you have *four minutes or more* to your assigned inbound time. The rule says you should be inbound as close as possible to your assigned time. When you arrive at the fix with less than *four minutes* to your inbound time, simply combine elements of a hold that will add up to the time you have left. For example: two minutes to go equals one 360-degree turn at standard rate; three minutes equals one pattern with 30-second legs. The only awkward situation is to arrive with one minute to go. There are two ways to handle that: either fly a standard rate 360 and depart inbound a minute late ("as close as possible") or call that last circuit a *five-minute* pattern and ask ATC if a one-and-a-half-minute holding leg is safe. At your airspeed, chances are it will be. Coming within a minute is generally pretty good, though.

The final approach segment is devoted to alignment and descent for landing. Positive course guidance must be provided, and once inside the final approach fix, you are required to maintain one nav receiver on the final approach facility without interruption. Therefore, if you are flying with only a single nav receiver, there can be no further retuning once inside the FAF. This segment begins at the final approach fix and ends at either the runway or the missed approach point. A final approach can terminate in one of two ways: a straight-in landing or a circle-to-land. One variant of the straight-in is called the side-step maneuver, in which two parallel runways aligned with a single final approach course allow a controller to clear you inbound on the course for one runway, but with a *clearance to land* on the parallel runway where sidestep minimums are published. It sounds like, "Clear for ILS runway.two four left, sidestep (or land) runway two four right." He also may add an "acknowledge" in there to ensure that you understand the nature of the clearance.

In general, though, a straight-in landing is pretty easy. It means the runway will be aligned within 30 degrees of the final approach course, the descent profile is not excessively steep, that the lowest published minimums will prevail and your odds of making it are pretty good in low weather. There is no radical maneuvering to be done and you simply transition from flying the panel to a look-up at the missed approach point. Got the runway environment (lights, markings, numbers)? Just continue the descent.

If the active runway to which the tower has cleared you is more than 30 degrees in alignment from the magnetic heading of the final approach course or the gradient is too steep, the higher minimums of the circle-to-land procedure apply. ATC will usually have stated that you can expect a circling procedure while you're on the approach, and it sounds like, "Cleared ILS runway two four, expect circle to land runway zero three, wind zero one zero at two five, gusting to three five." They're telling you there's a good reason for the circling procedure — the wind is pretty strong *across* runway 24. Such a clearance does *not* necessarily guarantee that you will have adequate visibility at the higher minimum descent altitude (MDA) to make it around to the other runway and land.

I wish there were a better term for what happens than the word "circle," because the last thing they expect you to do is to fly in a *circle*, and if you're new at this, it's misleading. As soon as you accept a clearance to circle, you must mentally shift gears to the higher MDA that applies to circling, and the minimums for ceiling and visibility may be pretty high. You may not proceed along the final approach course *any lower* than that MDA, and you may not descend from the MDA during the circling maneuver itself until you are in a position to land the airplane.

The circling maneuver is really a downwind or base leg to the final descent for landing. Until you are aligned on final and with the runway in sight, you must fly the "pattern" at the MDA. If you lose visual contact with the airport after you have acquired it visually, you must execute a missed approach. The rules state that you must take the *shortest* path to the landing runway from the final approach course, and there are no restrictions to that rule insofar as crossing the airport itself unless circling restrictions are published. A tower controller may stipulate a path based on his own traffic (many circling minimums are higher than VFR ceilings) that may *not* be the shortest path, however.

If you must miss the approach before you acquire visual contact, execute the missed approach associated with the straight-in procedure. If you have entered the circling maneuver and have visual contact, then you lose contact and must miss, always turn *toward* the airport. Then *climb*. The next step is to keep climbing.

You may request to use the straight-in landing anytime the tower stipulates a circle. There may even be situations where the crosswind is not sufficiently troublesome or the runway length is adequate to allow a downwind landing. Circling approaches in low weather, and especially at night, are something most pilots like to avoid. They can be pretty demanding, when you realize that you are transitioning to visual flying and maneuvering around the periphery of an airport, and with an absolute requirement for strict altitude control. A slight climb might put you into cloud and a missed approach; any descent cuts into your already meager 300 feet of obstacle clearance. A circling maneuver, then, divides your attention between the altimeter and the ground visual references, and that's no picnic.

Missed approaches have been widely advertised as the single maneuver most pilots perform worst of any instrument procedure; for the most part, the assessment is accurate. The missed approach segment is designed to be executed immediately upon reaching the missed approach point, the location of which varies with the type of approach. The initial missed approach procedure will always require a climb. Even if you can't find the chart, you forget the miss procedure, and you can't think of what to do — at *least* climb! There are three variations on the climb: straight ahead, left turn, and right turn. Most procedures further specify an altitude and a clearance limit; usually a hold at a navaid is the limit. Pilots who are blessed with two nav receivers normally tune the facility that defines the missed approach clearance limit or primary course guidance to that limit before reaching the final approach fix. Standard charts for approaches publish only the primary missed approach procedure; there may be other unpublished procedures as well.

Approaches are all designed to function around those particular segments, and it is actually less important for you to know exactly the dimensions of a particular segment than it is to understand the obstacle clearances allowed in each and to recognize the purpose for each segment. Approaches also come in two varieties: precision and nonprecision. The difference between them is easy to recognize; precision approaches offer glide path guidance, where nonprecision approaches offer only directional guidance to touchdown. "Precision" also connotes descents to lower ceiling and visibility minimums, and the standard straight-in ILS ceiling is 200 feet, with a half mile visibility. Obviously, not all ILSs offer those figures, but unless there's some external reason for raising the limits, the equipment is considered good enough to allow 200 feet and a half mile. Precision approach radar (PAR) is another form of precision approach, but it's offered mostly at military airports, and your opportunities to fly one are limited, usually to moments of extreme distress. Since a radar approach requires only two-way radio communication, it's a nice technique to know in the event of a failure of all your nav equipment in rotten weather and without the fuel to go elsewhere.

Nonprecision approaches include . . . well, all the other approaches — VORs, VOR/DME, NDB, surveillance radar, and even VHF direction finding. An ILS system without a glide slope operating — in other words, a localizer-only or a back-course localizer approach (or any localizer-type directional aid [LDA] or simplified directional facility [SDF] approach) — also falls into the category of nonprecision. Typical published minimums for a nonprecision approach are a ceiling of 400 feet and a visibility of one mile, but you'll find wide variations.

The most important piece of equipment when it comes to determining visibilities in minimums is the approach light system, and a good, operational light system can mean visibilities down to a half mile regardless of the type of approach, whereas inoperative lights or no lights at all may raise that figure to one mile.

Where precision approaches incorporate nav fixes in the form of marker beacons to provide distance-to-touchdown information, nonprecision approaches may use these aids to define a so-called visual descent point (VDP). The VDP has been added to nonprecision approaches to define a position from which a normal descent to a landing may be made provided the pilot has visual contact with the runway environment. VDPs are not considered part of the approach facility, and if they are inoperative, nothing changes. The VDP can be fixed by a DME distance or by a 75 MHz marker beacon, and a VASI (visual approach slope indicator) system of red and white lights provides the visual guidance that complements that nav fix.

The procedure for using VDPs goes like this: if you are at the MDA and can see the ground, don't descend until you reach the VDP, then begin a normal letdown (about 500 feet per minute or whatever the VASI lights indicate). If you fly *past* the VDP at the minimum descent altitude with no sighting of the ground or runway, all it says is that you can't land straight in on that approach, but you may be able to circle in the clear later on. Don't execute a missed approach at the VDP, though; the missed approach point is just where it normally would be: at the end of the appropriate time period or at the approach navaid (for an on-airport facility).

Nonprecision approaches are really nothing more than a route flown with a precise altitude, altimetry providing the height guidance in the absence of glide-path information from an electronic navaid. Think of it as a chance to fly over the airport as low as you can safely do so. Where a decision height marks the missed approach point on an ILS, a nonprecision approach is least precise in its definition of distance to touchdown and the missed approach point. Generally, you use a clock. As in tick tock. Begin timing from the final approach fix and fly inbound at a steady airspeed so as to maintain an estimated ground-speed; when time's up and you don't see an airport out the front window, execute the missed approach. Instead of descending to a decision height, you fly at a minimum descent altitude (MDA), and you may descend to that altitude at any reasonable rate as soon as you pass the final approach fix. Therefore, the final approach segment is not a gradual descent with vertical guidance but a descent to the minimum, a leveling off to maintain that minimum, and a slow cruise until your time runs out. Reduced to its essentials, it's a low pass.

Of the nonprecision approaches, the best and most precise are localizers, localizer-type directional aids, and SDFs because of their sensitivity to course deviations. Next come VORs and radar with DME distance information, then VORs naked, and last, NDBs. (There are some smart people who will tell you that NDB approaches are for emergencies. They are approved approaches; how good they are is for you to decide. Try them visually first before attempting one in low weather.)

Your information on approaches is limited to what's shown on the published approach charts, unless you dig further (into sources like the TERPs, local interagency letters of agreement, etc.), which few of us have an opportunity to do. Therefore, you ought to make the most of what the charts give you. Either an NOS or Jeppesen approach plate provides clear definitions of course, frequency of navaid, obstacles in the area, and missed approach procedure. They provide two views of the approach itself, so you can see what your path over the ground will look like as well as the altitude profile if you follow directions. Further, approach charts show you the principal navigation routes into the approach structure from the en route environment, both for planning purposes and for orientation. Here there is a significant difference between the two chart services, because Jeppesen may not incorporate the transitional navaids in the area covered by the chart, although the Jepps indicate the name of the navaid and its frequency with each feeder route or initial segment. The NOS solves the problem by ignoring distance scale so that outside a certain radius from the airport, the approach chart map may not represent proportional distances; the navaids are represented so that you can see them with respect to their bearing from the airport. Which you prefer is entirely up to you.

It's a lot more work, but the charts will be far more valuable to you if you've prepared them for use ahead of time. At the very least, you must review the likely approaches for your primary and alternate airports at some point prior to your arrival, for there will be no time to do it once you start descending into the terminal environment and aviating, navigating, and communicating at double the rate of the en route segment.

You should go back farther than that in your planning, however. Prior to departing, "fly" the approaches briefly in your head to evaluate their characteristics. Do you have all the nav receivers appropriate to the ground facilities to be used? What are the heights above touchdown for a straight-in, the height above the airport for a circling approach? Compare the available approaches to the destination forecast and determine to the extent that you are able the most likely approach you'll be using in order of precision.

If you care to make life really easy on yourself, there's much more you can do with the charts when they first arrive. Granted, the revision process to Jeppesen subscribers is not the favorite time of the week, but there are ways to take advantage of it to ease your job later. When you first get the charts, do at least this much for your home airport: highlight with a yellow felt marker the salient points for each procedure.

The three points you should have memorized at the FAF —
    How low? DH or MDA
    How long (or far)? clock time or DME fix at MAP
    Which way? the first part of the missed approach procedure
The frequency for the primary approach aid
The heading for alignment of the final approach course (*not* the runway!)
Minimum altitudes for each approach segment and stepdown fix
DHs and MDAs for both straight-ins and circling approaches, *for your category*
The time, at your expected groundspeed, from the final approach fix to the missed approach point
Tower frequency

Using charts correctly takes some study or you'll miss important information that may be there without your noticing. Before you use your new charts, even for training purposes, study the keys provided by both Jeppesen and the NOS for their products. If you can nail one down, attend Jeppesen's charts seminars with James Terpstra (no, the TERPs are *not* named after him) doing the narrative. Fail to study the chart symbology adequately and you'll make the mistake of one pilot who thought the letters "NA" under the ceiling and visibility minimums meant "not applicable." (They really mean approach "not authorized.")

Chart preparation means never having to say you're sorry you missed the note about the tower operation not being continuous and that's why you erupted from cloud 200 feet above touchdown to a totally dark airport after wondering why they wouldn't answer you. It means being forewarned that the neighboring airport's altimeter takes over and the decision height and visibility minimum have to be raised because the rule says you have to add five feet for each mile in excess of five miles the alternate altimeter is placed from the destination airport with the darkened tower. Gettin' to know an approach while you're already into it is the worst of all possible worlds, and since it's easy to avoid, there's almost no good reason to find yourself in such a spot.

There's fine print in them thar hills, and the little letters are hard to read in your average flickering map light.

Flying approaches is more than just the acquisition of information from all the various sources — ATC, the approach chart, the ATIS — it's a navigational chore as well. Approaches are intended to be as free of turns as the designers can make them, and radar vectors to an approach are ordinarily done with great consideration by most controllers to minimize extreme maneuvering. Almost without exception, a vector, even one from an arrival route that's directly opposite the final approach heading, will combine a series of gradual heading changes and descents designed to get you turned into a shallow intercept heading. Good controllers — or the ones who aren't busy — give you an easy angle of intercept, alert you to position with respect to the appropriate fix as they deliver your clearance or as they prepare to turn you over to your own navigation once you're established on course, and provide generous descent profiles that require no gymnastics to execute — most of the time. But every once in a great while, at some very busy period at a very busy terminal, you'll get something like "Cleared ILS approach, you're at the outer marker, call the tower, bye." It will happen while you're 2,000 feet higher than published for the segment you're on, and quite frankly, you're entirely justified in simply missing the approach right there and asking to go around again.

Most attitude flying in approaches concentrates on airspeed and altitude control. Approach procedures are, in the overwhelming majority, an arrow straight series of progressive descents to final. Thus, the skills you'll call upon most will be awareness of pitch attitude and power settings with respect to airspeed and rates of descent. Your panel scan should take in altimeter and rate of climb/descent (the VSI) more frequently, and you can help to discipline your altitude awareness by using call-outs, even when you're all by yourself alone. When approaching any altitude index, do as the airline crews do: call out "one thousand feet to go," "five hundred to go," "four hundred to go," and so on down to the minimum you want to hold. Same goes for final approach to DH or MDA. Sing out, captain.

Your pitch control scan will compare the bar-width nose-down horizon reference on your attitude indicator to the airspeed indicator, rate of descent, and finally, altimeter at varying rates depending upon how close you are to your reference altitude. As you approach your reference altitude, lead your asymptotic approach to the level-off with a little power, which will tend to raise the nose naturally without any pitch-trim correction. Part of learning approaches when you train is acquiring a set of power numbers that accomplish descents, level-offs, and airspeed adjustments. Five inches of manifold pressure or 500 rpm will generally provide a sufficient reduction of power to produce a 500-fpm descent, which is a nice pleasant, easy-to-handle rate for instrument approaches and will match the descent gradient of most three-degree final approach segments at typical lightplane groundspeeds. We'll get to electronic glide slope flying later; this just applies to altitude-change descents.

Directional control of the airplane marries attitude flying much more closely to navigation than it has been your experience to this point.

The best way to illustrate what I mean is to turn once again to the flight director. En route navigation is a slow, gentle monitoring of very slow rates of change and corrections made at a matching rate. It's a *low-gain* environment, meaning reduced sensitivity, high damping, low oscillation — get the picture? You are operating from a very *long* rope. On approaches, flight directors automatically tune themselves to *high* gain. Technical jargon aside, what happens is that the entire airplane seems to respond much more vigorously, more frequently and with control displacements that suddenly grow more solid and positive. Thus, if en route maneuvering were recorded in a book, you might find as few as three aileron inputs, say, per minute. On an approach, that might jump to thirty inputs per minute — ten times the number! *Approaches are characterized by more frequent control movements and corrections.* Some pilots say that approach flying is more precise, but that's a misreading of the situation.

All our navigation is accomplished in angular geometry, as we pointed out in Chapter I. You also found out that approach aids provide the same form of navigation geometry, and further, that on a localizer, for example, one dot of displacement on the CDI at the outer marker represents considerably more *linear* displacement — in other words, a greater number of *feet* from centerline — than the same dot at the middle marker. When we fly an approach, we're flying down into the increasingly narrow mouth of a geometric funnel. As you learned when we scanned the TERPs, the entire progression of approach segment boundaries becomes narrower as you draw closer to the airport. It is intended, therefore, that you fly ever closer to the ideal centerline that leads you to the runway.

Here is the least obvious single fact about approach navigation, and if you begin your IFR career understanding this single point, you'll make approach flying immeasurably easier for yourself.

The geometry of the approach guidance may duplicate the structure of the final approach segment. An ILS localizer is an electronic "funnel," widest at the entrance and narrowest and most precise in terms of linear dimensions at the landing end. It corresponds to the ideal approach structure.

The geometry of the approach guidance may be the reverse of the structure of the final approach segment. Any off-airport VOR, NDB, or omnidirectional aid provides a final approach course guidance geometry that becomes *wider*, in terms of linear displacement, at the landing end, than it is at the entry. Take a look at the figure on page 00 and see the difference.

Your problem, as the pilot and principal navigator of this miracle of aluminum, is to be able to visualize the navigational geometry in which you're flying and "flight direct" your airplane accordingly.

Within localizer-type directional guidance, your principal navigational indicator, the left-right needle on the CDI, will appear to become increasingly sensitive as you progress toward the airport. This is true, by the way, whether you fly the front or back course. (Back course flying is merely a matter of reversing the needle sensing — fly "away" from the needle.) If you had no directional guidance but that needle, you could not fly the approach comfortably — or perhaps, even safely. Why not? Because with no basis for comparison, the CDI needle on a localizer-type approach responds only to angular displacement, or *degrees* of off-course error. Your corrective maneuvers, however, *result in linear displacement of the airplane!* Pilots who "fly the needle" may do just fine at first, but as the airplane gets closer to the approach navaid, there is no cue as to how much correction is enough, and each correction becomes an overcorrection.

You could fly the needle, incidentally, if your airplane flew at an ever decreasing airspeed (well, groundspeed, actually, but we'll assume the two have a linear relationship for now), but that's impossible, so we have to figure out some basic measurement that will allow us to compare our correction maneuvers to some standard that won't change as we fly closer to the airport. Look around the panel. See anythng handy?

The DG.

And it turns out that flight directors fly perfect approaches by flying *heading* and electrically summing the heading value with the localizer course displacement, using the value for bank to provide the *rate* of heading change as the third term in the sum. If all terms cancel to zero, the airplane's *heading* is correct, and no turns are commanded. The exact dynamics of the flight director is less important than your understanding that ILS navigation is accomplished not by staring at the crossed needles — the CDI and glide slope — but by using the two instruments of *linear* displacement: the DG and the vertical speed indicator. Use the CDI indications to monitor your heading and to correct for drift, but nothing more. Use the glide slope needle to modify your rate of descent, but fly pitch and airspeed according to the rate needle, not the glide slope.

There's a school of instruction for flying the ILS that says that once you're established on final you should make heading corrections with the rudder pedals, and there are a thousand hangar stories about pilots "bicycling" down final in turbulence. There is *no* inherent advantage to rudder-only heading changes on a final approach, and the flight director certainly wouldn't fly it that way; it makes its heading corrections with bank. The actual reason behind rudder-only approaches is that the rudder is much less effective a control surface than the ailerons, and for a given amount of control displacement, in terms of pilot effort, the result is milder in terms of actual heading change. It was a way of preventing overcontrolling as the ILS course grew narrower and narrower, and the apparent sensitivity of CDI-pilot-controls, taken as a complete system, simply got out of tune to the approach geometry.

Pilots who fly approaches with the directional gyro for a primary directional aid work with a standard that undergoes no rate of change no matter where they are on the approach course and irrespective of the approach geometry. Like the flight director, they fly a final approach course like this:

First, establish the airplane on the course centerline (let's say you've just been turned in from a vector) and with a guess at the wind adjustment you'll need (you know which way it's blowing from the ATIS or from ATC). To do that, assuming a moderate intercept angle of 45 degrees or less, simply turn to your wind-corrected heading as soon as the CDI needle quivers with its first detection of the course; use a standard rate turn no matter how fast the needle moves toward center.

Three things can happen as you roll out onto the heading you've selected: you're on course or close to it with very little needle movement; your roll-out was a bit early and the needle is outside the turn and moving away; your roll-out was late and you've overshot.

All off-course corrections from here on are made in 5-degree heading changes, at least initially. If you're off course, you know on which side the course lies. Turn 5 degrees in that direction; a 5-degree bank held for a second or two will do the trick in most airplanes. Now you have a new reference heading.

At this point, you watch the CDI needle simply for its *motion*. Has your correction stopped its motion away from the centerline? If so, you have canceled the course error in your original estimated heading. You are now free to shift to a lower gear and make 2-degree heading corrections. Use rudder pedals if you insist, but little pulses of bank with the ailerons work a lot better.

If your correction has still not arrested the needle motion, add another 5-degree heading correction. *As soon as you stop the needle's motion, you have found your reference heading,* and your only job from here on is to adjust with 2-degree corrections until you get the airplane on the course centerline. Once you are established on centerline, return to the reference heading.

Now, it would be nice if all approaches went this way, with no wind changes on the way down to the airport. But you know and expect that you'll encounter changes in velocity and direction; that's what makes bad weather. There's no difference in the technique, however; your reference heading can be expected to change any time the needle starts to move again. Adjust the rate of reference heading change to match the rate of movement of the needle. When the needle once again ceases to move, you have a new reference heading and can adjust to centerline.

When you try it, you'll find that there is a world of difference in the combined chore of attitude flying and navigation on the approach. Very small corrections, made much more often, are a requirement of the sensitivity of the approach aid itself. While you were en route, your control corrections were adjusting the airplane within a very coarse geometry measured in nautical miles of width. On an ILS, each dot represents a few feet — quite a difference — and your airplane is still moving at half its cruise speed, so that deviations and course errors cover a lot of ground relative to the dimensions of the course itself. It's simply a matter of proportion.

The technique for flying the glide slope is like the localizer technique lying on its side. Remember, though, that the glide slope needle is *extremely* sensitive; it has to be in order to match the precision of the final approach descent path, where a single angular degree can change an airplane's landing performance markedly. There's also obstacle clearance to consider.

The sideways DG is the rate of climb/descent VSI needle, which establishes a steady rate of descent that will not change in sensitivity as you proceed down the approach course. The apparent sensitivity of the glide slope needle does change, of course, as the course narrows, just as the localizer's did. Jeppesen has gone to the trouble of providing on each approach plate for an ILS rates of descent corresponding to the glide path angle and various standard descent groundspeeds. They're in the lower left-hand corner of the approach plate, and they give you a leg up on estimating a rate that should stop the needle movement for you as you intercept the glide slope. The NOS prints a chart in the front of each bound Approach Procedure book.

As you arrive at the final approach fix, you may expect glide slope interception momentarily; the outer marker (FAF) is slightly inside the intersection of the glide path and the intermediate segment minimum altitude; usually the two are very close. Always glance at the altimeter at glide slope interception to ensure that the altitudes match; it's one way to prevent any possible chance of interception of the spurious glide slopes that are reflected upward at very high vertical angles.

As you near the FAF and outer marker, the glide slope needle will show some life. Lead the interception in one of two ways:

In a retractable-gear airplane, simply lower the gear; the added drag will almost always be exactly enough to start you descending. If the gear's already down, perform the power reduction as you would for a fixed-gear airplane.

In a fixed-gear airplane, lop off the descent power reduction you have filed in your head; about five inches of mp or 500 rpm for most airplanes does the trick.

There will be a few moments before the VSI needle settles on a stable value for the descent rate. It helps to have some initial reference — usually the attitude indicator is the best — to use until the rate of descent settles down. On most airplanes, about a bar width of nose-down will ease you through the few seconds until you can refer to the rate. Once you've established the rate you want, *then* watch the glide slope needle for movement. If there is no movement, your rate is just fine — touch *nothing*.

If the needle is moving upward slightly, add just a hair of nose-up attitude change, using the attitude indicator to make the change. It should be just barely perceptible. After a moment, this new nose-up attitude will affect airspeed and you'll have to adjust power to maintain your desired approach airspeed, which, after all, should remain on its reference.

If the glide slope needle is moving downward, add a fractional bar width of nose-down pitch. Watch the airspeed for the effect of the new attitude and adjust power accordingly.

Once you've stopped the glide slope needle from moving, you've found your reference descent rate for your airspeed. Any adjustment from this point on should be made to maneuver onto the center of the electronic glide path, then resume the reference descent rate. Glide path flying is a little trickier than localizer flying because your rate of descent translates through your groundspeed to become a gradient that matches the electronic glide path's downward slope of three degrees. Small changes in airspeed are okay so long as they don't appear to build in either direction. You'll hear arguments about whether to use elevator or throttle on a glide slope to adjust your descent rate, but they're about as helpful as arguments over whether to use rudder or aileron on the localizer. You use the *whole* airplane. You now have a complete picture of the invisible radio roadway you're flying along, and once you can see that and integrate the rates of change in the attitude flying instruments with the changing sensitivity of the navigation indicators on an approach, you're doing it the way a flight director would. Congratulations.

So much for ILSs, with their narrowing funnels. How should you fly the azimuth-only VOR approach, with its widening course, to a VOR that provides a two-degree-per-dot indication on your navigation indicator instead of the half-degree of the localizer. There is no glide slope, natch.

Since the VOR's geometry provides its greatest sensitivity close to the navaid — and far from the airport — you must fly such an approach with a different interpretation of the CDI. Overreacting to needle indications is far less of a danger here than it was on the ILS course, and on VOR approaches, the rudder-pedal heading change has no place and few adherents.

As you cross the VOR, the nav indicator will do its crazy number for a few seconds and the to-from indicator will reverse. Start your clock when it does and turn to the inbound heading. Also set the OBS to the inbound value. The actual order of these first acts is not terribly important, and since most pilots don't look at a clock but simply start a stopwatch, you can be turning to your final approach heading. Once you're on that heading, retune the OBS. At this point, the VOR needle may be off the scale. Pretty alarming, huh? Not really, because your awareness of the approach aid's geometry tells you that within a few hundred feet of the station, an off-course linear error of 20 feet may be enough to uncenter the needle.

Wait for the needle to come off the peg, but if it hasn't after about 15 seconds, make a 10-degree heading change *toward* the needle. There is no great hurry about VOR approaches, so you'd be better off waiting than overreacting initially to the CDI needle. If you still haven't acquired the course after a 10-degree heading change, either you have a very strong wind or something is very wrong, so start checking the inbound course and the OBS setting to ensure you're right.

Chances are that by now the needle has begun to swing toward center, and you'll notice as time goes by that for a course that's even roughly approximate to your initial heading, the needle movement will begin to slow down as you proceed away from the station even if you make no correction. Naturally, you understand this phenomenon of apparent decreasing sensitivity and are prepared to take steps.

First, do what you have to do to center that needle. Use 5-degree-bank aileron for *decreasing periods of time* until the needle is centered and not moving. Your DG is now probably within 5 degrees of a good reference heading for the remainder of the approach. Notice how the world has different dimensions on a VOR approach? Make your corrections with authority, using 10 degrees of bank and rounding off the reference heading to the nearest 10 degrees, so that you fly a heading within 5 degrees on either side. VOR approaches require *increasing* attention to needle deviation because the course is getting wider the farther you fly toward touchdown. The corrections do not match that changing rate, though. If anything, your corrections of 10-degree bank may be held longer the farther you fly.

The difference between ILS and off-airport VOR approaches can be illuminated best this way: if you picture the flight of an airplane down a final approach course as a wavy line of very tiny correcting turns to the right and left of the centerline, on an ILS that wavy course will eventually fly off the electronic pathway. On the other hand, the wavy course line on a VOR approach could actually become more divergent and still present the pilot with good on-course indications.

There is yet a third case, the on-airport VOR, in which the sensitivity of the course geometry increases somewhat as you near the airport, but in this instance, the supposed enhanced accuracy does you no real good. The narrowing dimensions of the final approach course do not lead you to your goal — the runway — but to a very sharp VOR antenna, and unless you're the spot-landing champion of the world, you can't put an airplane down on one of those. On-airport VORs are a pain because they offer you no final approach fix (unless the approach incorporates a DME or a second facility, that is) and the nav facility itself forms the *missed approach point,* not the point of landing. The higher minimums associated with on-airport VOR approaches reflect these disadvantages.

To keep your mind occupied, as if you had nothing else to do, the world of instrument flying has invented a cockpit procedure that applies to all navigation but particularly to approaches; and of approaches, particularly nonprecision approaches. This cockpit procedure is called the Six Ts, and it goes like this: Turn, Time, Tune, Throttle, Talk, and Track (the sixth is optional since it's more or less understood). You might call this little mnemonic system a Fix Procedure, since that's where it really applies. The Six Ts represent a mental (I hope) checklist that helps you to proceed in an orderly fashion through the list of chores you have to accomplish at an approach fix, and in their proper order of importance.

It tells you first to turn to the inbound heading which you *think* will put you on the correct approach course. Start the clock. On an ILS, you can save your day by keeping track of timing — many pilots don't bother because the missed approach point has a definition (the decision height) — and in the event the glide slope flies off to glide slope heaven, you continue merrily on your way, observing the slightly higher minimum descent altitude of the localizer-only approach. (By the way, if you forget the clock, use the middle marker to miss; at least you'll be safe. No marker? Bet you forgot to turn the MB receiver on. Either way, miss right now!)

You've turned and timed. "Tune" covers a quick check of the nav receivers for the correct frequency and switch positions; navaid ident "on" and at low volume so you can monitor it all the way down the approach as an audio backup to the flags on your CDI; marker beacon on and tested; ADF tuned to the compass locator (which can sit in for the marker beacon if the MB transmitter is out of service); and, on a VOR course, the OBS tuned to the proper inbound radial.

A word here on localizers. When you tune a localizer frequency on your nav receiver, you'll notice that the OBS ceases to function. The localizer presents simple course information and doesn't measure phase difference and translate into radials the way the VOR does. That leaves you with a useless OBS knob — or does it? Most pilots either use it to dial in the inbound course heading or to set the missed approach outbound course if it's different from runway heading. I think it's better to do *something* with the bearing scale than to leave it on some old course value that might confuse you if you glance at it during a busy moment.

"Throttle" really means "altitude" and "airspeed," and even if no altitude change is called for, it's your cue to check altitude quickly. Usually, a final approach fix means a descent, so reduce power and start down.

"Talk" comes last and for good reason. Most final approach fixes are miles from the airport and take at least two or three minutes for you to make the trip to the runway. Which means you've got plenty of time to tell the anxious tower controller that you're a-comin' in. He already knows that, of course, since approach told him and nowadays he may even have his own radar.

"Track" really means "get on with it." Start navigating for the appropriate approach geometry and center the needles.

Now get ready for a rush, because one of the finest moments in this life is the sight of a runway the very second you break out of a cloud on course and on final. It's enough to make a romantic out of you. It did me.

IFR Notebook

The Six Ts of IFR Procedures

Turn — heading
Time — clock
Twist (or Tune) — navaids: frequency and course
Throttle — power/altitude
Talk — communicate to ATC
Track — fly on course

Runway Visual Range Values (RVR)

RVR is a system that provides continuous automatic measurement of horizontal visibility near the runway from the approach end. It is used as the basis for visibility criteria when high intensity runway lights spaced no more than 200 feet apart are operative and the appropriate runway markings for instrument conditions are in place. RVR values represent *feet* rather than fractions of a mile, rounded off to even hundreds or even thousands.

RVR values vs. ground visibility in statute miles

RVR	visibility
1600	¼
2400	½
3200	⅝
4000	¾
4500	⅞
5000	1
6000	1¼

Aircraft Categories for Approaches

Approach categories for different aircraft recognize that airplanes of different weights and landing performance require entirely different volumes of airspace in which to perform approach maneuvers. Airspeeds that are used to categorize airplanes for approaches are based on 1.3 times the stall speed in the landing configuration at maximum gross landing weight. The lowest category for any airplane is determined by *either* the weight or the airspeed criteria, whichever puts it in the higher category. For example, an airplane that meets the category B airspeed maximum but exceeds the weight maximum should be categorized in accordance with the higher weight maximum.

	airspeed	weight
Category A	less than 91 knots	less than 30,001 pounds
Category B	91 to 120 knots	30,001 to 60,000 pounds
Category C	121 to 140 knots	60,001 to 150,000 pounds
Category D	141 to 166 knots	150,001 or more
Category E	166 knots or more	*any weight*

Instrument Landing Systems (ILS)

Components:
1. directional guidance — localizer
2. glide path guidance — glide slope
3. range or distance guidance (and timing reference) — marker beacon
4. visual information for *alignment, roll,* and *distance* — light system
   alignment — centerline of light system, "rabbit"
   roll reference — crossbar assembly light system
   distance — crossbars, touchdown zone lights

Localizers:
40 channels, VHF 108.1 MHz to 111.95 MHz
Front Course: angular width between 3 and 6 degrees to provide a linear width of about 700 feet at the threshold; accuracy is ± 1 degree
Back Course: REVERSE NEEDLE SENSING
  use only with a published approach procedure
  dimensions of course vary with transmitter placement with respect to runway of primary (front course) use
Ident: Morse code letter "I" (two dots) followed by three-letter facility ident
Area of coverage — course information outside these fan-shaped areas is invalid:

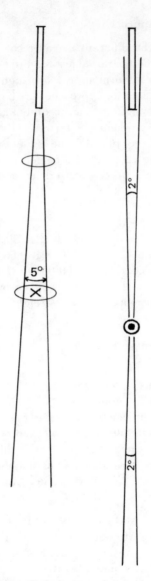

The way you navigate varies with the approach. The nav needle of the CDI behaves quite differently on an ILS (very high sensitivity) and an off-airport VOR (much lower sensitivity). Be aware of the different geometries of final approach courses and navigate accordingly. The linear width of the ILS course (left) decreases steadily from the FAF at the outer marker to the flare point.

(Right) This VOR (also its own FAF) provides a course line that gets wider, in linear dimension, as you fly final approach to a landing.

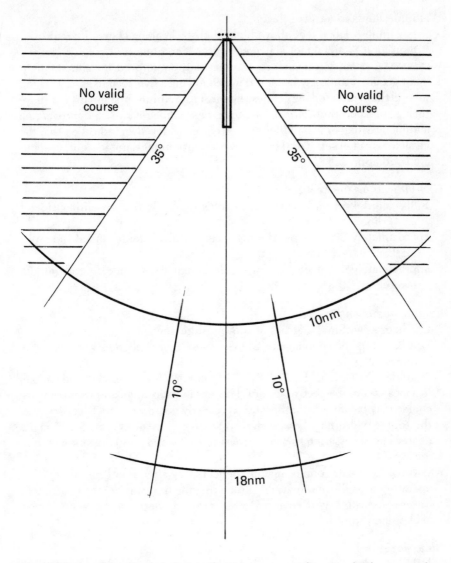

Normal limits of localizer coverage; the same area applies to a back course when provided.

altitude limits to localizer beam:
   low limit — 1,000 feet above highest obstacle along course line
   high limit — 4,500 feet above elevation of antenna site
WARNING: Certain localizers are automatically monitored for course alignment accuracy and may be shut down following a five-second alert warning
tone if their radiated signal deviates from a minimum standard for more
than a specified time. *Always monitor ident while on an ILS approach;* an
audible warning or loss of ident should be considered an indication that the
localizer is out of service. If the runway environment is not in sight, execute
a missed approach.

CDI indications on localizer:
   Full-scale deflection normally 5 degrees total, 2½ degrees either side of
      centerline
   approximately 300 feet per dot displacement from centerline at the outer
      marker
   approximately 100 feet per dot displacement from centerline at middle
      marker

Runway alignment:
   Localizer — within 3 degrees of runway alignment
   Localizer-type directional aid (LDA) — more than 3 degrees variance from
      runway alignment
   Simplified directional facility (SDF) — a *nonprecision* approach aid similar
   to a localizer in its electronics but different in its location and geometry; its
   course may be offset from the runway (usually not more than 3 degrees) and
   the course width may be adjusted wider than a localizer's, usually 6 or 12
   degrees as necessary to provide "maximum flyability and optimum course
   quality."
WARNING: Localizer courses may be deflected by aircraft or ground vehicles
   operating near the antenna site. Protecting the transmitter area is a control
   tower responsibility. If a control tower is closed, there may be no assurance
   of alignment protection.

Glide slopes:
   UHF, 40 channels, paired with localizer frequencies and automatically
      tuned by selecting localizer frequency; 329.15 MHz to 335.00 MHz
   normal glide path — 3 degrees above the horizontal, greater with approval
      of Flight Standards
   Useful range — normally 10 NM
   vertical width of beam — 1.4 degrees
   Optimum threshold crossing height — 50 feet
      Maximum crossing height — 60 feet

Usable height above runway, or flare point — 18 to 27 feet
CDI indications on glide slope:
   Full-scale deflection normally 1.4 degrees total, 0.7 degree above and
   below ideal glide path
   approximately 50 feet per dot displacement from glide path at the outer
   marker
   approximately 8 feet per dot displacement from glide path at middle
   marker
WARNING: Glide slopes are provided only for matched localizer courses for
one approach to a runway. Back-course glide-slope information is not valid
unless specified in the approach plate. In addition, false needle indications
may be received at very high vertical angles. Cross check interception of
glide path for correct altitude to avoid descending on a glide path that is
higher than the intended course.

Marker beacons:
   single frequency — 75 MHz; all markers transmit on this frequency
   power — 3 watts or less
   signal dimensions — ellipse 1,000 feet above antenna
   2,400 feet wide
   4,200 feet long
Outer marker ident — 400 Hz tone; two dashes per second (– –); location —
   approximate lowest glide-path interception, about 5 miles from the
   airport
Middle marker ident — 1,300 MHz tone; 95 dot-dashes per minute (•–); location — 3,500 feet from touchdown, typical
      200 feet above touchdown, roughly equal to decision height for a preci-
         sion approach
Other uses: final approach fix for back course, inner markers for Category II
   approach decision height
   Marker beacon light indications (panel)
   white — fan or bone marker
   blue — outer marker
   amber or yellow — middle marker

Compass locators:
   Primarily an aid to orientation, these NDB-like aids operate through the
ADF receiver on 200 to 415 kHz and act as a backup to the marker beacon by
indicating needle reversal at station passage, and as an orienting aid by providing bearing to their associated marker beacon locations prior to final approach
segment.

The Rules
When an instrument letdown to an airport is necessary because of any limi-

tation to inflight visibility, you must use a standard instrument approach procedure published for that airport in FAR Part 97, which means using an approach published by any approved instrument approach chart source, such as the NOS or Jeppesen. The rule does *not* require that you have the approach plate in possession, only that you use the approved procedure.

You may not land an aircraft using a Part 97 procedure if the visibility is below the minimum published with that procedure. If the minimums include a ceiling criterion, the visibility minimum still applies but the ceiling minimum must be added to the field elevation and that value observed as the MDA (for nonprecision approaches) or the DH (for precision approaches).

If the approach specifies "No Procedure Turn" or "NoPT" or "FINAL," you may not execute a procedure turn unless you tell ATC when you receive your approach clearance.

You may not descend below the prescribed minimum descent altitude or continue the approach below the decision height unless your aircraft is in a position to make a normal approach to the runway on which you intend to land *and* you can clearly see the approach threshold, the approach lights, or other *markings* that identify the approach *end* of the runway. If you have continued the approach past the MDA or DH and then subsequently lose visual contact with the approach end or any of its associated lights or markings, you must immediately execute the missed approach procedure.

Departures

There is no applicable visibility minimum for Part 91 operations. However, pilots are expected to comply with SID (Standard Instrument Departure) procedures where they are published or to notify ATC if unable to comply with the SID, either because the pilot lacks the charts or *because the airplane may not perform adequately to meet minimum climb standards*. The NOS lists SIDs in a separate section of approach plate books, while Jeppesen publishes the SIDs in chart form, usually indexed to precede STARs and approach procedures, organized by city name.

All pilots in *all* operations should consider prior to departure the terrain in the airport area and any obstructions either charted or uncharted, and (1) use a SID if published, (2) determine whether terrain or obstruction clearance can be maintained visually, or (3) at airports where no procedure has been published, determine what action will be necessary to assure a safe departure and then fly accordingly. Normally, a missed approach procedure will indicate the most likely avenue of exit for obstruction clearance; where none is available, use a sectional chart to determine your own departure procedure.

Terminal Instrument Procedures (TERPs) Obstacle Clearance Criteria
(look out below!)

Minimum safe altitudes — en route structure
  1,000 feet of obstacle clearance *for emergency use*:
    Minimum sector altitudes: within 25 NM of nav facility, if nav facility
      located within 25 NM of airport, otherwise a radius of 30 NM around
      airport
    Emergency safe altitudes: military procedure, 100 NM from nav facility,
      but 2,000 feet in mountains
Initial Approach Segment Using Straight Course and Arcs
  Area no standard length but no more than 50 NM
  6 miles on each side of course centerline, with
    primary area 4 miles each side of course
  Obstacle clearance
    1,000 feet in primary area
    500 feet in secondary area
Initial Approach Segment Using Procedure Turn
  Area Normally 10 NM from fix to maneuvering reference
    otherwise as published in profile view of approach
    5 NM where only Category A aircraft will operate
  Obstacle clearance
    1,000 feet in primary area
    500 feet in secondary area

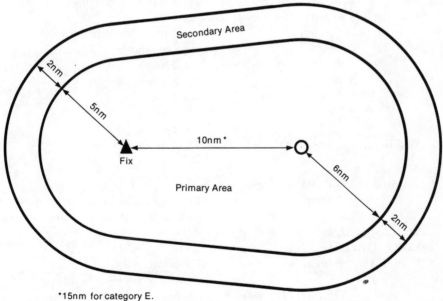

*15nm for category E.

Obstacle Clearance Chart

Intermediate Approach Segments Using Straight Courses
  Area not less than 5 miles nor more than 15 NM
    optimum 10 NM
    width determined by drawing a line between outer boundary of final
      approach course segment and initial approach segment; primary
      area is defined the same way
  Obstacle clearance
    primary area   minimum 500 feet
    secondary area   500 feet tapering to 0 feet
Intermediate Approach Segments Using Procedure Turns or Arcs
  Area
    length — similar to straight course but curved in shape
    width — for arc 6 NM each side of course with primary area 4 miles
      each side of course
        — for procedure turn, refer to initial approach PT depiction
  Obstacle Clearance
    primary area   500 feet minimum
    secondary area   500 feet tapering to 0 feet
Final Approach Segment
  Visual Descent Point Obstacle Clearance Area
  no obstacles: at an angle 1 to 1½ degrees *lower* than glide path and within a
      fan-shaped area 10 degrees each side of the runway centerline
  Circling Approach Obstacle Clearance

Category	Circling Radius (from any run-way boundary)
A	1.3 miles
B	1.5 miles
C	1.7 miles
D	2.3 miles
E	4.5 miles

Within areas defined by these radii, *300 feet* of obstacle clearance

Approaches
On-Airport VOR   (no final approach fix)   straight-in *only*

Primary area	2 miles wide at facility to 6 miles wide at 10 NM from facility
Obstacle clearance	300 feet

VOR, VOR/DME   (with final approach fix)   straight-in *only*

VOR/FAF	*Primary area*	2 miles at facility to 5 miles wide 30 miles from facility
	Obstacle clearance	250 feet
VOR/DME	*Primary area*	4 miles each side of course

Obstacle clearance	500 feet

NDB, On Airport   (no FAF)

Primary area	2.5 miles wide at facility to 6 miles wide at 10 miles from facility
Obstacle clearance	350 feet

NDB, with FAF   straight-in *only*

Area	2.5 miles wide at facility to 5 miles wide at 15 miles from facility
Obstacle clearance	300 feet

ILS Systems   straight-in

Area	length: 50,000 feet from a point 200 feet outside the runway threshold width: from 1,000 feet total from a point 200 feet beyond the threshold to 16,000 feet wide at 50,000 foot border
Obstacle clearance	defined by a surface with inner 10,000-foot section and outer 40,000-foot section, and with a gradient of 34:1 for the inner portion of the typical 3-degree glide path and 29.5:1 for the outer section slope

Missed Approach Obstacle Clearance

Area	width of final approach segment at MAP, expanding to width of initial approach segment (usually 12 miles total) at a point 15 NM from missed approach point
Obstacle clearance	defined by a surface sloping upward at a gradient of 40:1, equivalent to approximately 230 feet per minute at 90 knots groundspeed

NOTE: a curving missed approach surface is usually slightly wider to accommodate maneuvering space for the turn

How to Convert SID Climb Rates into Feet Per Minute

Groundspeed/climb factor (multiply times *feet per mile*)

60 knots	×	1.0
70 knots	×	1.2 or fpmile $+ \dfrac{\text{fpmile}}{5}$
80 knots	×	1.3 fpmile $+ \dfrac{\text{fpmile}}{3} =$ answer is climb rate in feet per *minute*
90 knots	+	1.5 or fpmile $+ \dfrac{\text{fpmile}}{2}$
100 knots	+	1.67 or fpmile $+ \dfrac{2 \times \text{fpmile}}{3}$
120 knots	+	2.0 or 2 × fpmile

CHAPTER IX

Deviations

This final chapter is about the moments in instrument flying when you may actually get excited. Another book might call this chapter "Emergencies," but I've heard too many knowledgeable people agree that what we're really talking about here is simply procedures you don't use very often. These procedures are published, and their very existence shows that they've been well thought out; hence the argument that in IFR there are no "emergencies," merely inconvenience or momentary excitement.

It is a simple fact that these procedures are also the ones in which pilots have the least practice and are therefore least current. You'll have to apply an element of judgment, therefore, when you think about deviations. In every case, they apply to command of the airplane, not management of the airspace, and that puts you squarely in charge of the situation and responsible for its outcome, for good or bad. There has never been a procedure written for abnormal circumstances with punitive intent. Aside from a check ride requirement, your skill at steering by a wet compass alone will impress no one once you begin flying and you encounter a failure of panel instruments. The minute anything goes wrong with your airplane or with the flight, you will be amazed at the resources that will be brought to bear, the spirit of cooperation of all concerned, and the relief when you finally land your airplane safely.

You want to know what a real emergency is? It's what you create when you fail to notice that your microphone button is stuck and you block a frequency. (When it happens, you have to use the microphone jack as a switch if you forgot your spare — shame. To transmit, insert the jack; to receive, pull it out.)

An emergency is failing to wait for an acknowledgment when you decide to cancel IFR. For that matter, it's any device you use to *disappear* from positive control without telling anybody. When you do that, you completely screw up cubic *miles* of airspace for everybody until ATC has had a chance to sort things out and determine that you are *down* somewhere. Then the fun starts.

Phone calls to airports ("Anybody seen a %?*$# single? He's lost on an IFR flight plan") and search-and-rescue procedures follow. Once, an airplane crashed on approach to an uncontrolled airport. Because so many pilots fail to close IFR flight plans after arrival, can you really blame the local guy for deciding *not* to call out the St. Bernards? It lay there for hours, without help. Anyone who has ever failed to close out properly contributes to situations like that one. Closing an IFR flight plan recently became mandatory, and failure to do so is a violation.

An emergency is accepting visual clearances and the implied acceptance of traffic separation without actually having the traffic or without calling back if you lose the traffic, once acquired. An emergency is being rusty — or anything but current — and doing nothing about it. In short, a real emergency is a dangerous pilot. There has never been a rule or procedure written that can cope with that threat. All the world can do is to try to act *after* the harm is done. That's an emergency.

I was coming back from somewhere down south one evening with my girl friend in the right seat and a business associate hanging over us from the back seat like a big, wet dog, fascinated by everything that was going on — he knows no fear, the fool — when the weather suddenly deteriorated radically in our destination terminal area. It was about seven o'clock in the evening, and when you take off on the second leg of a 1,000-mile-total trip, with the sun gone out and the fatigue creeping up on you, you sometimes don't check the weather for a destination 400 miles ahead of you as carefully as you should.

I had filed no alternate, for I've always figured you might as well file "Chicago" no matter where you are since your real *alternate will be the nearest airport with adequate weather. There's the flight plan and then there's the real world. . . .*

Anyway, I started the letdown into approach control's bailiwick and there ensued the usual first-contact-with-approach litany: "Second floor gent's underwear third floor bath shop, wind one eight zero at twelve, temperature 98.6, ceiling 100 overcast, visibility one quarter, what are your intentions?" That was the first I knew just how bad things had gotten, since my rudimentary weather checks of my destination area had made it clear this was no really acute storm situation but just one of those dank, rainy, foggy evenings that had made rush hour a mess and would probably add up to a lot of late dinners all over town.

Well, what were my intentions? To pull over and park while I figured this thing out — if I could have had my way about it. But we know how foolish that is, having beaten the subject to death in every accident clinic and classroom and tsk-tsk-ing the other guys who've made the statistics, smirking over their incompetence. A whole lot of past was coming home to roost.

What really confused me was that here he was telling me that the airport's reported weather was below the minimums for the published approach, yet he was asking me what I wanted to do. (". . . your intentions?") *I'd almost rather he'd said, "Hey, fella, the field is closed, the weather is junk, we're all coming unwrapped, two airplanes have missed before you, and you better go away because I want to go home." But no (". . . your intentions?").*

So I said to them, "Oooh, rog-ah, I'd like to have a look." To "take a look" – it was an airline captain phrase I'd heard somewhere, and it grew from that little eternal area of confusion over what constitutes minimums. If the tower is reporting zero-zero but the touchdown zone is 500 and a half, can I land? I knew the answer, had heard it time after time, but suppose I did land tonight? What would happen? Suppose I landed and bent something? Insurance canceled. Payments for the rest of my life. . . .

Then the tower saved me further embarrassment by saying just at the right moment: "Confirm you're going to have a look?" It had that we-are-now-officially-compiling-the-dossier-for-this-impending-crash tone to it.

I suddenly realized I was on tape.

We went to the big airport, where it was still 250 and a half, shot the approach, landed, rented cars, and got home three hours late. It occurs to me as I tell this that it was the one *occasion in five years of flying that I had to go to an alternate because the weather was too low. And that's what bothered me most.*

That's a real story, told by a very good pilot. And it illustrates an important point about IFR flying. First, the worst times are the moments when you don't know what to do. Second, the rules don't answer every question. For Part 91 operations, it is true that you, as pilot in command, determine whether you have sufficient visibility to make a safe descent to landing at the missed approach point. You can, if you like, shoot and reshoot the approach until some chance gust of wind blows the cloud away temporarily and lets you in. It's become one of the most uncomfortable loopholes in the regulations because airplanes don't come with visibility-measuring equipment. Airlines and Part 135 operations have no decision to make: if the *reported* weather is below minimums, they don't shoot the approach. Period. You have to make up your own mind, however; drop as many hints as you like, but nobody in the tower or at approach control is going to make the decision for you. They'll simply report the weather with about as much emotion as they would the six o'clock news.

The pressure that compels you to continue in the face of one person's opinion of the weather may come from any source. If you fly for pay, arriving with the goods in time may force you to press on. Your car is parked at the destination airport. You're due somewhere early tomorrow morning and it's too late to cancel; a deviation will ruin everything. Plans. Promises. Convenience. Think about it.

The important point to that story was simply that over the long term, your plans aren't changed that often because of the weather. IFR flying delivers you on time, on target, with a percentage somewhere in the high 90s. In our own minds, we forget the approaches to an airport that unexpectedly turns VFR, those times when IFR almost isn't needed. For some reason, pilots forget how generous the percentages are when they're faced with an approach to below-minimums conditions. It becomes a challenge with a risk that's completely disproportionate to the reward, assuming everything ends safely. So the answer to the inevitable question is yes, you can shoot approaches into airports that are essentially closed. My advice is to use those occasions for practice. Everyone should shoot an approach in actual conditions to a miss once in a while to rekindle the basic notion that all approaches *miss* unless proved innocent. You need to be reminded periodically that every descent down an ILS *doesn't* terminate with the "rabbit" in sight and a clearance to land, and too few pilots have experienced a missed approach within the period of required currency for basic IFR. So fly into low conditions if you want to. Just don't feel so compelled to land.

I sometimes wish weather forecasts were accompanied by some kind of rating system – like the movies' "G" and "R" – that would tell a pilot the quality of the information on which the forecast is based. In this particular misadventure, I already knew that the route forecast for an early Sunday morning would be based on the least reliable information: it was over a long stretch of desert with no weather observation stations except at my departure and destination terminals. At this hour, there wouldn't even be so much as a single pilot report to confirm or clarify what somebody a thousand miles away had figured out from the upper atmosphere picture.

Flight Service got my attention right away with the word that cloud cover extended over most of the route. By staying down at the MEA, I could stay out of it, though I knew I'd be out of radar contact on one stretch and communications would be iffy. The winter temperatures aloft told me the chance of ice in the cloud itself wasn't worth playing with; I filed the MEA.

After departure, I took a VFR segment to avoid a climb over a short high-MEA airway segment crossing some mountains, and ATC obliged. From the closer vantage nearing the ridgeline, I could look ahead into the desert and see what lay before me. There was a lower layer – looked thin, though – that nobody had reported (obviously, since nobody'd been through here yet today) and I passed that along to the controller, who sounded kind of sleepy but friendly.

At the handoff, center and I exchanged the expected loss-of-radar-contact formalities, as I tried to make out the weather between the layers ahead. I was still at the MEA.

I still had communication but no radar contact when the first drops hit the windshield . . . and stayed there. For perhaps thirty seconds, I watched, fascinated, as the clear ice formed on the windshield as a kind of transparent-plastic rippled material, almost like one of those shower doors. For all that was on the windshield, there seemed to be very little on the wings; at least there was very little visible. Should I do something now, or wait a minute? Maybe this would just go away. It sure wasn't in the forecast.

Without my noticing it, the com radio had gradually begun to generate static, almost as if the squelch were fading off. No question, it was time to talk to ATC about this situation and get turned around. I realized all of a sudden that ice accumulation on the wings would be greatest where I couldn't see it — on the under surface. The airplane was beginning to feel sluggish.

No answer to the first call. I was still concentrating on heading, navigation, altitude, flying as I'd been drilled to fly so many times, and as I think about it now in retrospect, the idea never dawned on me that I already knew there was no traffic out here. If there had been, this would have been reported!

Within something like three minutes — I'm sure that was the actual time now, though it seemed more like fifteen or twenty in the compressed moments of worry — the airplane was in serious trouble. Then I thought I heard an answer, very faintly through the squelch noise, and whether it was center or not, it was all I needed. Somebody was on the frequency, and through the antenna's coating of ice, I could at least receive him. The odds said he'd hear me.

Carefully and very gradually, I nursed the airplane around in a 180 and held attitude. I was a thousand feet below the MEA now, but familiar enough with the route to be pretty sure of generous terrain clearance. As soon as I had it turned and level again, I called in the blind [meaning without established radio contact] to advise anybody that I could no longer maintain assigned altitude and that I had made a turn back to reverse course.

I landed at the first airport I could find in the clear and climbed down to look at the load of ice that had accumulated on the belly antennas and the rest of the airframe. One antenna had broken but not separated, and the rest seemed okay. I don't know, and I certainly don't care, whether ATC heard that first call or the ones I made later — neither one of us noticed the clock — but the first time I heard an answer from him, it was obvious that he'd heard me sufficiently on prior call and I had been right — I was the only target in that mess this morning.

After two hours of thinking about it, I filed back to my home base and parked.

That came from a pilot with several thousand hours of flight time but very few experiences like that one. Still, his decision and the events that led up to it illuminate how doggedly we'll stick to the drill in any situation, thinking that there's safety in the rituals of IFR — heading, altitude, MEA, communication — all meaningless in freezing rain. Finally forced into a decision for which there is no actual rule, he made one up — and that's the whole point. No controller or government agency would criticize him for allowing the airplane to descend below the MEA or for turning around without a clearance to do so.

The common denominators to all deviations are that they are necessary, they are simple, and they make good common sense. It's the part of IFR flying that calls for judgment, and it's the least tangible part of the whole enterprise. Every instructor wishes he or she could teach judgment, but there's no way. Every piece of rule making tries to substitute for judgment; nobody can write the perfect set of rules, though, and the *least* helpful regulations and paragraphs in the *AIM* and doctrine in the training manuals are those that allude, in a vague way, to judgment. About the best way I can think of to touch on a subject that's essentially untouchable is to relate some real stories that happened to real people. Keep flying IFR long enough and something similar to these will eventually happen to you. Oh, I don't mean you'll necessarily fly into clear ice or a foggy airport. I mean you'll have a moment when you honestly won't know what to do. At that moment, IFR will seem less like a system designed to increase your chance of arriving and more like a regulatory trap that led you into a pickle. The rules will seem to surround you and yet give you no guidance. It is the least attractive aspect of learning to be an instrument pilot.

The important thing for you to realize is that the system was not invented to lead you away in manacles at the drop of a problem. ATC is a mute witness to your meanderings from the ideal, but in the event of genuine difficulty, their role is one of assistance, not enforcement. If you misinterpret this as license to endanger or even inconvenience others, you do so at your own risk, however. A pilot who frequently broke off a low approach in IFR weather conditions to land at a nearby non-IFR field that was also below VFR once too often, then claimed an emergency when he was called on it, was thus in a very tough bargaining position.

It's hard to give much advice about the bad moments that arise out of unforecast weather and embedded thunderstorms and the like. They simply happen, and they are part of the risks of this game. If you don't accept that from the start, you are in for a rude shock.

Some deviations can be thought out in advance, though, and dealt with in training. Various losses of equipment are one example. Take them in order of "aviate, navigate, and communicate."

Loss of a panel attitude instrument is jarring because it knocks the props out of the reflexive habits of attitude flying, and that can be a distraction. Even professional air crews get very little practice flying with an instrument missing except at regular check rides.

If you lose the vacuum source, you will be left with the static pressure instruments and the electrical ones.

Lose the electrical system and you will have the vacuum-powered group and the static pressure instruments.

Loss of the electrical system ordinarily means you'll have to modify your navigation a little too, since you may want to cut down significantly on the battery drain. If it happened to me, I'd land as soon as I could.

Many pilots consider loss of radios alone to be no emergency at all. An instructor I know, who prefers to remain anonymous, has this to say about a *confirmed* loss of communication: "It is secretly my biggest thrill; loss of coms alone is the most wonderful favor that could come my way because I know exactly what procedures to use and one thing's guaranteed: no delays."

Even loss of an engine imposes very little uncertainty, although it may raise your heart rate. At least you'll know what to do. On a multiengine airplane, feather the bad one and find a nearby place to land — don't prolong the flight unless you're already close to the destination. Get it down. With only one engine — and this solution has been beaten to death — you have no decisions to make. Establish the configuration that will give you minimum sink rate (*not* best glide), tighten the straps, and get ATC to help you find the best place for the landing.

When you lose panel instruments, you should tell the controller the nature of the problem and your intentions. Some pilots carry devices to blank out inoperative instruments during a failure, but since you can reasonably expect the occasions for using them to be rare, you can rely on a makeshift equivalent. That lightweight approach chart paper will adhere to an instrument face with the simple addition of a little spit. A three-by-five card may be wedged into the face of the thing under the panel shield. Anything that tends to "flag" the instrument will serve as a reminder to disregard its indications so that you don't burden yourself with conscious reminders to avoid watching a bad attitude indicator, say.

The controller will switch to "no-gyro" procedures if you report loss of vacuum instruments or, in any case, the DG. Thus, you may effectively eliminate flying by the wet compass alone as a standard emergency procedure — and a good thing too, since it's an awful instrument to steer an airplane by. No-gyro turns are calculated by the controller. Instead of giving magnetic heading assignments, the controller observes your path across the radarscope, determines your magnetic course, then, with a watch, assigns heading changes this way:

"Turn left (or right) . . . (*they* time the number of seconds at standard rate for the heading change they want) . . . stop turn."

That's all there is to it. Controllers are supposed to stay current in these no-gyro procedures, and I've had some ask if they could practice on me. If you want to try one, you might inquire sometime if somebody wants to give a practice no-gyro. In nice weather, it's no great inconvenience to you, and it will give you a chance to gain confidence in the procedure.

You'll learn to operate with a wet compass during your IFR training, and a lot of inspectors still require ability with the instrument in maneuvers. Since it's a skill that takes plenty of practice to stay good at, don't count on it in your bag of tricks later on, though. If you find yourself stuck in cloud and steering with a compass, just estimate your time for the turn and aim for standard rates. If you come within 10 degrees on the roll-out, you're an expert. Just make minor corrections from there, and try to minimize the number of turns.

Let's say for some weird reason you lose your nav receivers or indicators but you can communicate okay. Obviously, your en route segment shouldn't be too bad, since radar vectors will get you around just fine. But how do you shoot an approach? If the weather is not too awful, you can usually get a radar surveillance approach even if a procedure for it isn't published in the charts. Let's take the situation one step further, though, and put you in a real bind: the whole East Coast is down to 200 feet or less in a quarter mile and fog and you just had a navigation failure so you couldn't shoot an ILS if your life depended on it — and it does. You can talk just fine, though, and if you can get help fast, your hour's fuel remaining may be enough to save the situation. What do you do?

Look for a military airport or some facility with precision approach radar. Usually, PAR is associated with military airports because the DOD guys make approaches that way. The FAA's experimental facility called NAFEC at Atlantic City, New Jersey, is another well-known PAR approach, but that won't do you much good inbound to San Diego. Get on the horn and ask for the nearest PAR facility as early as you can.

Here's how PARs work: two radars working simultaneously stand in for the localizer and the glide slope. One radar provides course or azimuth information, and the other provides height or glide path data. A controller on the ground has both scopes in a single console and talks the airplane down — just like the movies.

The surveillance radar will get you lined up generally into the initial approach and give you a lost-communication procedure which is exactly the equivalent of what you'd do if the flags came on during an ILS approach — you execute a miss. Difference is, they *tell* you the miss procedure, so you have to write it down or remember it. Think of it this way: the controller's voice is your navaid; lose it and you miss the approach. Also, at a certain point, after they turn you over to the final controller (which is sort of the same as the final approach fix), the controller will say something like "acknowledge no further transmissions." From that point on, the controller does *all* the talking. If you so much as touch the mike button, you've wrecked everything, so don't.

On a PAR, you fly airspeed, DG, and the VSI needle, using the attitude indicator to integrate and generally keep your wings level. The controller will give corrections like:

"Turn left two degrees. On course. On glide path. Turn right five degrees. On course. Slightly above glide path. Correcting. On glide path. Two miles from touchdown."

Like that. You'll get a 30-second warning before intercepting the glide path, but they won't tell you the decision height unless you ask.

You may also obtain direction-finding assistance from any ATC or FSS that's equipped with the DF receiver and scope, but in order to get an IFR approach in actual conditions, you have to first declare an emergency. DF equipment is inherently less accurate, and that's why there's a limit on its use for routine approaches. Bakersfield, California, is an example of one airport where there is no radar coverage but DF guidance instead, and the specialists there get lots of practice.

Here's how DFing works: the specialist or controller assigns you a frequency to work on for controlling purposes. First, you reestablish communication. Next, they'll briefly check their equipment and your course. The receiver provides them with a "strobe" or line of light that tells the azimuth from the antenna to your airplane. You are asked to key your mike — don't say anything into it, just key it — for a short period. By measuring the azimuth change and the time against your airspeed, they can get an estimate of your distance from the station. You tell them when they ask you what your compass says and they establish your position and course. With timed turns, they'll get you oriented very gradually to a large approach course with generous buffers. They'll also give you altitude assignments down to minimums for their published procedure. DFers assume you are *not* IFR rated unless they know otherwise. On the few occasions I've tried it at Bakersfield, they took me right down the white runway stripe, so the procedure can be very accurate in good hands.

Too many pilots think a two-way communications failure means anytime they can't transmit. It's less important to be able to transmit than to receive, and it turns out there is a trump card transmitter aboard and perhaps as many as three auxiliary receivers for audio information aside from the communications receiver.

The ace-in-the-hole transmitter is the transponder. Think about it. Once they figure out that you can't talk with words but can receive, ATC can, if they have to, ask you to acknowledge with certain codes or to answer simple questions with different codes for each answer. You may not be able to answer "What are your intentions?" but you can handle something like "Can you land VFR? Ident if affirmative."

If the communication receiver acts as if it's gone belly up, monitor it anyway and start listening for help on other receivers. Turn up the volume on the VOR to which you are navigating. Listen to an NDB that's nearby and has voice capability. And believe it or not, there's audio on a DME too. Turn the volume up and listen for the identifier and any transmissions.

Don't be in a great hurry to tune 121.5 just because something's gone wrong. Too many pilots have it backward and assume that if you have an emergency you first switch to 121.5 because that's the emergency channel, after all. Nonsense. If you have established radio contact, report a Pan or Mayday on the frequency you're *already using*. 121.5 is for those times when you haven't been able to raise anybody on another frequency. There is a mistaken impression that the whole world is listening to 121.5 all the time; if that's true, they're sure turning the volume down. I've tuned in to a pilot who's been in trouble for half an hour on the frequency with nary a whisper of response from the ground. You'll get much better action on a busy frequency for your initial report of a Mayday or Pan. Once you have their attention, they'll tell you when to switch if they want you to.

When communication is completely lost and there's not a prayer of regaining it, you enter into the realm of the lost-com procedure, which is nothing more than a deviation from the third task — communicate. You may assume that as soon as ATC stops hearing you, you will gain an extraordinary amount of attention. Having confirmed a lost-com problem, you also tune 7700, which rings bells in the radar room and causes blossoms to appear around your airplane's symbol on the radarscope. Having triggered the alarm with 7700, you may turn to 7600 after a minute or so. (There are times to be exact in your clock timing, and others when it's less important. This is one of the latter.)

Now you can follow the rules that are outlined briefly for quick reference in the IFR Notebook for this chapter.

Your route should be direct to the nearest VFR airport if you're already VFR or if you encounter a *reasonable* opportunity to *land* VFR somewhere along the way. (Oh, and when you do, please call ATC and tell them you're safe. Like your mother, they worry about you.) This doesn't mean you have to dive through any old hole and plummet down into a canyon. It means that if a reasonable opportunity presents itself to land at an airport you can see and to make the descent, approach, and landing visually, there is no good reason for continuing IFR. It's hard to imagine anyone *not* doing exactly that, but someone must have or they wouldn't have had to write a rule.

Otherwise, if you have to complete the trip IFR, follow the route in the clearance you got most recently before the failure of communications. If the failure occurs while you're on radar vectors, fly direct to the fix you were given in the vector clearance. You'll recall that once ATC takes you off your own navigation and assigns you a magnetic heading to fly with the words "vectors" buried in there somewhere, they have taken over the navigation role. Loss of communication means that navigation reverts back to you. *Always* ensure that a vector clearance from a controller includes some identifiable fix in IFR conditions (in VFR, they may say "vectors to the traffic pattern," but that's different), such as "This'll be vectors to the ILS final approach course," etc. The operative word after "vectors" is "to."

Obviously, this situation can cause a real problem, depending on the fix. In the example of a vector clearance limit to the ILS final approach course, you'll have a real problem navigating direct if your panel lacks an ADF and the fix is the compass locator. There are moments like that when you just have to invent something sensible, and there are a thousand variables. If your transponder still shows you a reply light, *somebody* probably has you on radar and can keep other traffic away. That leaves terrain to worry about first. Don't assume that the heading at which they left you when the com failed will necessarily be the right one to follow straight home. You begin to see why people who've been through something like this have studied the terrain beforehand or keep their sectionals handy. Jeppesen's terrain contours on the area charts are a big help here.

In any case, just because the rule says *fly direct* doesn't mean you should stop thinking.

Next in order, if neither of the other two situations applied, is your *expected route* according to the *most recent clearance*. The word "expected" is in the glossary of ATC to take care of lost-com procedures, and any time there's a clearance limit short of the destination airport, the word *expect* shows up with respect to two and perhaps three items: time, route, and altitude. You'll find situations wherein ATC just won't give you an expect-further time because the system hasn't given *them* one. Off the record, they'll tell you "we *expect* to see you at the airport," in other words, just keep truckin'. As things stand, though, you have to play it by ear if you lose com with no "expects" in your pocket.

Last, if none of the above situations apply, you fly the route you filed in the flight plan. You should find yourself in such a bind only when you have taken off from an uncontrolled field into IFR conditions and cannot establish radio communications with anyone. Just fly your route. (Makes the notion of a com check before departure sound kind of important, though, doesn't it? At least try unicom or another airplane first.)

The rules for routing are pretty straightforward. What altitude do you fly at?

On any *segment*, which means, essentially, departure or en route, since approach altitude is in the next procedure, you fly the *highest* of the following: the last assigned, the MEA, or the "expected" altitude. Here's how that works in practice.

Let's say you had a clearance to climb and maintain 8,000 feet on a westbound airway, with an added "expect 12,000 at Bloss intersection." On the climb-out, you lose com. First, climb to 8,000 feet. I can't imagine that they'd assign you an altitude *lower* than the MEA, but check anyway. (If the MEA is higher, continue climbing till you reach it and maintain that for this segment.)

Along your route, and prior to reaching Bloss, there's a short section with an MEA of 11,000 feet. Climb to 11,000 *even though that's usually an eastbound altitude.* As soon as you depart the airway segment with that higher MEA, you must return to 8,000 feet again. At Bloss, climb to 12,000.

Approaches are different, and they have their own set of special lost-communication procedures that prescribe both time and altitude.

One rule covers the situation if you are holding somewhere en route. Your problem is to figure out when to leave the fix and continue the flight to the approach. You should have one of two "expects": an expect *further* clearance time, or an expect *approach* clearance time.

If you have an EFC time, leave the holding fix at that time and proceed by the route that fits the rules to the most direct approach procedure unless one has already been assigned in a clearance. Usually, you'll have a route that terminates at some fix that defines the approach you'll use (an "implied" approach); if more than one approach begins at that fix, use the most direct, as ATC will protect all of them.

If you are ahead of your ETA but have no holding instructions, hold at the fix where the approach begins (initial or intermediate) until the ETA, then begin the approach. Suddenly, the reason for devoting loving attention to your flight-plan en route time (which actually produces the ETA figure when it's added to your *actual* departure time) becomes apparent. I don't know why pilots are so generous about adding ten minutes for the wife and kids to their en route estimates. All you do is increase the chances that you'll have to hold at the IAF. There are no penalties for being late in a radar environment, so why pad the en route estimate?

If you have received an expect approach clearance time in the course of any clearance, that becomes your lost-com time at which you are expected to begin the descent from the appropriate en route *altitude* for the approach itself. Don't begin the descent prior to that time. With no EAC time in any clearance prior to the com failure, you can begin the descent at the ETA implied in your flight plan, as amended by you if you're in a nonradar environment.

Begin the descent at the initial approach fix.

You got it right.

The way the rule is written, you can be assigned a flight level and you must maintain that for the en route portion of the flight if it satisfies the highest altitude rule. You'd have to lose all that altitude between the initial approach fix and the final approach segment. It's obvious that no airplane in the world can lose that much on, say, an NDB approach where the initial approach fix is the beacon and the initial segment is the outbound procedure, but that's the way the rule is written. In a case like that, the best way to lose the altitude is to fly a holding pattern around the fix until you reach the published altitude for the initial approach segment. You have to lose the altitude somewhere, and that's probably your safest bet.

The reason why it's the safest bet is because the final rule for lost-com procedures in approaches says that if for reasons of timing you must hold prior to commencing an approach, you should do so in the approach fix holding pattern, if one is published, or if none is published, hold on the procedure turn side of the final approach course, using the FAF approach aid for the holding pattern fix. Just use the pattern to lose altitude as you would in a stacking procedure.

You can be heartened by evidence that these procedures are rarely used and that when they are, they seem to work well. For one thing, your attention sure picks up when something goes wrong, and you're likely to be more alert than you've ever been to considerations of terrain clearance and procedure. It's the routine of comforting ATC handling that makes you complacent, and there can be situations where that's more dangerous than these so-called emergencies.

If any of this worries you, remember the night of the great blackout in New York City when an entire terminal area had a huge failure of the *ground* system. The combination of backup power and communication, coupled with a good share of common sense, got every airplane down safely. If pilots and controllers working together were able to get through that one, individual instances of uncooperative machinery seem a whole lot less threatening.

IFR Notebook

Weather

"To the extent possible, controllers will issue pertinent information on weather . . . areas and assist pilots in avoiding such areas *when requested."*

How to request a weather-avoidance deviation:

1. Forget the propaganda; *ask* if there's weather on your route!
2. State the number of miles and magnetic heading of your requested detour; the number of miles means lateral displacement off the airway centerline.
3. Can you navigate a parallel course on your own? (RNAV equipment often can provide automatic parallel) *if not,*
4. Request a new airway route or direct navigation to avoid the weather area.
5. Request a change of altitude (if you really think it'll help).
6. Request radar vectors around the affected area (bear in mind that vectors place a high work load on the controller, who may already be pretty busy).
7. Make the request as early as possible; unless you're in cloud, you should be able to see isolated thunderstorm activity well ahead of the point where you're in it. Listen for reports of icing from other aircraft.

Emergency Locator Transmitters

Replace the batteries —

, after more than *one hour's* cumulative transmitting time
or, when the battery has reached 50 percent of its useful life

How to Report ELT Signals
Tell any Flight Service Station the position and time when:
signals first heard
signals lost
signals at maximum strength
your flight *altitude* and the *frequency* (almost always 121.5, although
   ELTs also transmit on 243.0 MHz)

Communications Failure — Suspected
   1. Determine if it's an ARTCC failure; switch to backup may take about
one minute, so keep monitoring if you have two radios, then
   2. If you've just been transferred, recontact the transferring controller
   3. If you lost com after contact was established, try
      a) any other known ATC frequency in that sector (listed under the lit-
          tle telephones on Jepp charts), including terminals, then
      b) the next sector along your route, then
      c) any Flight Service Station
   4. Listen on navaids
      a) VORs
      b) NDBs (use the ADF "receiver" setting)
      c) DMEs
   then,
   5. Squawk 7700 for one minute, 7600 for 15 minutes and repeat

NOTE: Not all radar facilities are equipped with automatic alarm for radar alert
to a transponder emergency code 7600 unless already in radar contact and
control. That's why you squawk 7700 first. Use Mode 3/A (no altitude framing
pulses) for emergency squawks.

Communications Failure — Confirmed and Unable to Receive
   If you are VFR or encounter VFR —
      LAND!
   If you are IFR in IFR conditions,
      Route: last clearance
                 if on vectors, to where they said they were vectoring you
                 expected route
                 route filed

EN ROUTE   *Altitude:*   the highest of —
                             last assigned
                             MEA (even if it's the wrong way)
                             "expect"

APPROACH
>If you are holding with holding instructions when failure occurs,
>>— leave *holding* fix at expect *further* clearance time
>or , if you have an expect *approach* clearance time,
>>— leave *holding* fix to *arrive* at the *initial approach fix* at the EAC
>>time

>With no holding instructions *and* the airplane ahead of ETA,
>>— hold at initial approach fix until the ETA, then commence approach
>>— which approach? the most direct; they're all protected by ATC

>When to descend from en route to approach
>>If you have an expect approach clearance time, at the EAC time;
>>If not, at your ETA

>Where to descend from en route to the approach
>>At the initial approach fix

Where to hold if you have to
>>In the approach fix holding pattern, if one is depicted, or
>>On the fix, using the procedure turn side;
>>If neither depicted, you may not hold.

Magnetic Compass Errors for Partial Panel Maneuvers
>*Turning Errors*
>>— vary with angle of bank; 15- to 18-degree bank
>>— lag to the north
>>— lead to the south
>>— are minimal at east and west
>turns to the north: lead the roll-out heading by
>>your latitude plus half the angle of bank
>turns to the south: overshoot the roll-out heading by
>>your latitude minus half the angle of bank

Acceleration Errors — on easterly or westerly headings *only*
>>Accelerate north, decelerate south (ANDS)

How to make an emergency report:
>Ident three times
>Type airplane
>Position or estimate (and say which)   OVER
>>(Pause to ensure continued com contact)

Heading/magnetic or true (and say which)
Altitude
Fuel remaining   OVER
       (Pause to ensure continued com contact)
Nature of distress
Intentions
Assistance desired

DO NOT spew out all elements of the report at once. It'll take nearly three minutes to say, and at the end of that, you may hear "Could you say that all over again please? I dropped my pencil."

Rule for Political Hijackings to Certain Listed Countries
    Wait for the next election. We will probably have embassies at all of them. And squawk 7500.

A Final Word

Instrument flying is your chance to be as professional as you want to be about your flying. You may never get to fly an airliner or a military high-performance airplane or satisfy those fantasies that everyone secretly harbors, but by applying yourself to your training and supplementing your talents with some study and determination, you can earn a rating that allows you to fly in the same ATC system that these pros use. One of the most satisfying rewards to IFR flying is the chance to work with controllers who will be at least as experienced as you are and, the odds say, probably with many more hours logged at the radarscope than you have with an airplane. It's an environment that tends to push you a little, but that's good. You'll find that the skill of those around you will rub off on you with time, and you'll get better.

Nobody expects you to charge out there and do everything perfectly from the beginning. The things you have to learn can't be taught in books, the foremost of these being confidence. Confidence doesn't come with reading about it; it comes from going out and doing it yourself. IFR flying will bring out the best in you. At the conclusion of every flight, you'll think of something you could have done slightly better, and seldom do you make it from departure to approach without learning something.

If everything you've read here has slighted a single aspect of IFR flying, I'd guess it would have to be the sheer emotion of it all. IFR gives you the kind of success you'll savor more than most. Granted, the world can't peer into the cockpit to see how well you're doing in there, so it's a private sort of reward. To compensate, there are those moments when you'll sense a spirit of teamwork with the groundbound half of the flight — the ATC people — the feeling that you're working together as equals. When that happens, you'll find a sense of comfort with your airplane that no other form of aviation can satisfy.

Go out and fly. Try real weather and sample the variety that the sky has to offer. Sure, a clear day in sparkling sunshine can be joyous beyond measure, but once you get your instrument rating, you'll find yourself the odd person out in a crowd, secretly wishing it would cloud over in time for your trip. Clear skies are fine, but the inside of a cloud is like another world. It'll give you a whole new appreciation for your airplane too. At night, with the panel lights glowing and each instrument and radio doing its job, it'll be putting out its all for you, the whole machine working in harmony in a way you never realized it could.

But don't take my word for it. You have to be there yourself. Instrument flying is one of the few things left in life that's really worth doing.

INDEX

Accessories, 36
Accident reporting, 15
Advisory Circulars, 14–15, 35
Aeromedical Institute, 35–36
Ailerons, 54
Air Defense Identification Zone (ADIZ),
    130, 140
Airframes
    cracks in, 29
    fatigue, 27
    measuring life of, 29
Airman's Information Manual (AIM), 14, 108,
    120, 140, 144
Airmets, 156
Airplane performance, 135, 136
    procedures, 160
Airport advisory area, 130
Airport directories, 16
Airport/Facility Directory, 41, 136, 138
Airport traffic area, 130
Air-powered instruments, 48–49
Air pressure, 49
Air pumps, 49
Air route traffic control centers (ARTCCs),
    112–14
Airspeed
    altitude and, 136
    attitude indicator and, 64
    clearance, 124–25
    climbs and, 64–65
    controllers and, 124
    drag and, 62–63
    as an element of the flight plan, 123–24
    en route descent and, 65–66
    graph of, 136
    for holding, 103
    in navigation, 74
    pitch and, 59, 60
    procedures, 168, 180
    as a trim function, 59
    turns and, 63
Airspeed indicators, 45, 59, 60, 64–67
    navigation and, 92
Air temperature, 135, 155
Air temperature indicators, 53
Air Traffic Control Handbook, 108
Air traffic control radar beacon system
    (ATCRBS), 111–12
Air traffic control (ATC) system, 12–13, 14,
    20, 26, 58, 72, 104–34, 212

altitude designations, 59, 106–8, 117, 123,
    125
in clearances, 119–23
communication within, 104, 106–9, 114,
    122, 126–27
    gaps in, 113
    hidden, 117
    importance of, 127
    loss of, 107, 121
    responsibility for, 113–14
computers in, 112, 113, 114, 117, 120, 121
deviations and, 195, 200, 203, 204, 206
flight plans and, 139, 140, 143, 144, 146,
    150, 156
fuel requirements, 109
holding and, 94–96
IFR Notebook, 128–34
intimidating effect of, xi
learning basic machinery of, 108–9
light signals, 131
navigation and, 86–88, 90–94, 107
pilot's request privileges, 105
procedures and, 159, 160, 162, 164, 166,
    168–70, 175, 178
radar in, 110–18, 122, 124
recording of conversations, 14
reports required by, 132
traffic conditions and, 109
training in, 19
weather conditions and, 109, 112–13, 120,
    123
wind conditions and, 124, 133
See also Controlled airspace
Air traffic control towers, 12–13
Airway, defined, 130
Alert notices, 156
Aigorithms, 74
Alphanumerics, 114, 116
Alternators, 31, 32, 33, 49, 70
Altimeters, 39, 45, 49, 67, 131, 175
    as a barometer built around a different
        scale, 49
    climbs and, 64, 65
    encoding, requirements regarding, 71, 72
    en route descent and, 65
    nonreporting, 60
    pitch and, 59, 60, 62
    requirements for testing, 39–40
    settings for, 92
    turns and, 63, 64

TWEB, 141

Ultrahigh frequency (UHF), 117
Uncontrolled airspace, 110
United States Department of Transportation, 12, 15, 159
United States Government Printing Office, 14–15, 143
United States Standard for Terminal Instrument Procedures (TERPs), 108–9, 159, 160, 161, 165, 173, 174, 176, 190–92

Vacuum pumps, 31–32
Valves, 30, 31, 50
Vectors off published routes, 125–26
Vertical speed indicator (VSI), 2, 45, 49, 50, 61, 66, 67, 70, 175, 179–80
  climbs and, 64–65
  instantaneous, 59
  pitch and, 59–62
  turns and, 63
Very high frequency omni range system (VOR), 77, 80, 82–88, 91–94, 98–99, 102, 173, 203–4
  advantages of, 83–84
  crab-angle detection, 93
  flight plans and, 139–40
  mysterious characteristic of, 85
  procedures, 180–82, 186
  required checks for, 39
  testing, 40–42
VFR flights, 2, 4, 16, 25, 131
  altitude and, 120, 123
  changing to or from, 107, 123
  choosing over IFR flights, 118–19
  clearances for, 120, 123
  compared to IFR flights, 43, 46, 47
  dead reckoning in, 74–75
  deviations and, 198, 200, 204
  navigation, 74–75, 102
  in preparing for IFR flights, 18–19
  the weather and, 135, 139
VHF emergency frequency, 107
VHF receivers, 14

Vibrations, 27–29
  load stress compared to, 28
Visibilities, 118–19, 133
  estimating, 128–29
Visual descent point (VDP), 172
Visual orientation, 1–3, 6–7
VOR, *see* Very high frequency omni range system
VORTACs, 8, 71, 91, 92, 94

Weather, ix, 135
  ATC system and, 109, 112–13, 120, 123
  deviations and, 197, 198, 208
  flights plans and, 135–37, 140, 141, 143, 149, 151, 154–57
  VFR flights and, 135, 139
Weather-briefing sheets, 141
Weather radar, 112–13, 156
Wet compass, *see* Magnetic compass
Wilkinson, Stephan, 141
Wind
  approaches and, 178
  ATC system and, 124, 133
  forecasts of, 151, 155
  in navigation, 74–77, 80, 82, 86, 88, 91, 93, 103
    crosswind, 77–79, 91
    downwind, 91
    headwind, 80
Wind triangle, 74
Wings
  flex of, 28
  fuel tanks in, 28
  keeping level, 55
  load on, 27, 66
World War II, ix–x
Written Test Guide, 14, 15

Yaw, 43, 45–46, 51, 54, 56, 57, 65
  change in, 44, 45
  rate of turn and, 58
  stability and, 55
Yaw axes, 45, 46, 58
Yaw trim, inclinometers and, 65